COMPUTER SCIENCE, TECHNOLOGY AND APPLICATIONS

AN INTRODUCTION TO APPROACHES AND MODERN APPLICATIONS WITH ENSEMBLE LEARNING

Computer Science, Technology and Applications

Additional books and e-books in this series can be found on Nova's website under the Series tab.

COMPUTER SCIENCE, TECHNOLOGY AND APPLICATIONS

AN INTRODUCTION TO APPROACHES AND MODERN APPLICATIONS WITH ENSEMBLE LEARNING

YI-TUNG CHAN
EDITOR

nova
science publishers
New York

Library of Congress Cataloging-in-Publication Data

ISBN: 978-1-53618-680-2

Published by Nova Science Publishers, Inc. † New York

Contents

PREFACE

From the successful application of deep learning (DL) in AlphaGo in 2012 to the recent advances in edge computing, artificial intelligence (AI) has continued to develop over the years. In the face of the current sweeping trend of AI, ensemble learning (EL) is expected to be further applied to DL and AI for developing higher-level ensemble systems in the future. Moreover, it could become an important step for achieving "The Master Algorithm" proposed by Prof. Pedro Domingos. In light of this, EL will continue to make a significant contribution to future development. This book aims to provide practical insights from experts in computer science and data science as well as other fields to help readers in establishing the basic concepts of AI and machine learning (ML), thus enabling them to solve problems in their fields by utilizing EL techniques.

EL is an AI and ML tool that is simple and practical, the concept of which is used in our everyday life and work. In simple terms, EL is similar to a parliamentary or jury system, where people from various industries and backgrounds express their views on a particular topic. This approach covers different levels of knowledge in various fields. Assuming that everyone has fair judgment and a certain understanding of the topic, combining everyone's ideas can reduce errors in individual decision making, ultimately achieving collective decision making. EL uses the philosophical notion that "two heads are better than one" as its core concept. In AI and ML, the "heads" that work together to find a solution refer to such differing and diverse minds as learners, agents, or detectors, who classify data and make decisions. According to existing theories, an ensemble system that combines multiple accurate, yet diverse, base classifiers has been proven to enhance the overall classification accuracy.

Furthermore, EL is straightforward and practical, and knowledge of EL can help us to obtain relatively high-accuracy results and establish a robust ensemble system when facing classification problems in real life. The techniques and concepts related to EL have been commonly used in various international ML and data science competitions to achieve good results. In recent years, EL has been adopted by most of the winners for international data mining competitions, such as the competitions hosted by Kaggle and the "KDD Cup" organized by the Association for Computing Machinery (ACM). For example, a team led by Prof. Chih-Jen Lin, Prof. Shou-De Lin, and Prof. Hsuan-Tien Lin from National Taiwan University obtained precise predictions by integrating different models established by each group within the team; further, they have won numerous prizes at the ACM KDD Cup in the past six years. Moreover, extreme gradient boosting (XGBoost) has been one of the most commonly used EL techniques in Kaggle competitions since its introduction in 2014. EL has been proven to be effective and applicable for solving real-life problems in data analysis.

The purpose of this book is to provide insights into EL for readers not majoring in computer science or related subjects, introduce the latest development and applications of EL; in particular, share its practical applications in various fields. Accordingly, this book intends to present theoretical parts relating to mathematics and computing in a simple and concise manner. The examples and practical use of EL have been used to explain methods that utilize EL to solve readers' issues in their fields, which demonstrates the essence of EL for practical applications. While many AI and ML books are available on the market, most require a certain level of mathematical and ML knowledge. Complicated theories of mathematics and computation may be intimidating for people without a background in computer science and engineering, such as biological and medical researchers. It would be unfortunate if they were to miss the opportunity to use EL as a practical tool to solve data analysis problems at hand. Moreover, EL is usually introduced in the later or advanced chapters of AI and ML books. Beginners in ML, or readers without a technical background, are likely to be frustrated by mathematical or technical terms that only appear occasionally in the book or be anxious about complicated mathematical and computational theories related to classification algorithms. It would be regrettable if they were intimidated, and therefore, missed the opportunity to learn and use EL. From a practical perspective, existing classification techniques, such as decision trees with the C4.5 algorithm, support vector machines, and neural networks are now relatively mature and have been proven to be effective. For readers without a technical background, it is not essential to understand the complicated mathematical and computational theories behind the above techniques. Instead, it is recommended to grasp the logic and meaning of parameters in these classification algorithms and directly conduct tests using EL. Learning through practice can help readers to establish computational thinking. It is the best approach to learning EL, ML, AI, and DL.

This book provides insights into EL from worldwide experts and scholars in various fields. This book extensively introduces and discusses the application of EL in various fields and the current and future research directions of its novel applications. It also reviews some of the more popular areas in which EL has received widespread attention in recent years in the ML and AI. Each chapter opens with an introduction to ML and EL techniques, and then, analyzes the applications of EL in different fields, such as signal and image processing, medical care, education, geology, and agriculture. More than two experts and scholars in related fields acted as reviewers for the peer review of each chapter. It is hoped that these applications in various fields can inspire readers to use EL in practice.

ACKNOWLEDGMENTS

We would like to acknowledge the individuals who contributed to this effort. First and foremost, we would like to express our sincere thanks to the contributors of the chapters for reporting their research and for their time and promptness.

Furthermore, we highly appreciate the willingness of the reviewers to take the time to complete this review. The high standard of a book can only be maintained with excellent and dedicated reviews of the chapters by such leading experts. We know that good reviewers are busy with other professional commitments, and once again, we would like to express our cordial appreciation of your dedicated efforts to support our activities.

The reviewers of this book are as follows: Prof. Jen-Chun Lee (Department of Communication Engineering, National Kaohsiung University of Science and Technology), Prof. Chao-Lung Chou (National Defense University, Chung Cheng Institute of Technology, Department of Computer Science and Information Engineering, Taiwan), Prof. Chih-Yi Chiu (Department of Computer Science and Information Engineering, National Chiayi University, Taiwan), Prof. Akihiro Yamashita (Department of Computer Science, National Institute of Technology, Tokyo College, Japan), Prof. Barbara Motyl (Industrial and Information Engineering, University of Udine, Italy), Prof. Domenico Speranza (Design Methods and

Tools for Industrial Engineering, University of Cassino, Italy), Prof. Wen-Cheng Lai (Department of Electrical Engineering, National Penghu University of Science and Technology, Taiwan), Prof. Hsin-Te Wu (Department of Computer Science and Information Engineering, National Ilan University, Taiwan), Prof. Chanadra Whiting (Department of Wellness, Broward College, US), Prof. Anu Saini (Department of Computer Science Engineering, G. B. Pant Engineering College, India), Dean and Distinguished Professor Shu-Kun Hsu (College of Earth Sciences, Earth Sciences National Central University, Taiwan), Research Geophysicist Victoria Langenheim (Geology, Minerals, Energy, and Geophysics Science Center, US), Prof. Le Nguyen Quoc Khanh (Medical Humanities Research Cluster, School of Humanities, Nanyang Technological University, Singapore), Dr. Kuo Ching Yuan (Department of Emergency and Critical Care Medicine, Taipei Medical University Hospital, Taiwan), Prof. Shuenn-Jyi Wang (National Defense University, Chung Cheng Institute of Technology, Department of Computer Science and Information Engineering, Taiwan), and Prof. Ko-Chin Chang (Department of Electrical and Electronic Engineering, Chung Cheng Institute of Technology, National Defense University, Taiwan).

Next, we would like to thank Nova for publishing this book and for their advice and patience. We believe that this book provides an important contribution to the EL community. We hope that this book will serve as a motivation for continued research and development in EL.

Finally, I dedicate this book to my family, with love and gratitude, to my mother Yu-Chih Tu, father Te-Pin Chan, wife Pei-Ju Lin, daughter Chien-Chien Chan, and son Chien-Yi Chan. Thank you all for making me a better person.

Funding: This work was supported by the Ministry of Science and Technology of Taiwan [grant numbers MOST 107-2218-E-012-001-MY2, MOST 108-2221-E-012-005, and MOST 109-2221-E-012-003-MY3].

In: An Introduction to Approaches … ISBN: 978-1-53618-680-2
Editor: Yi-Tung Chan

Chapter 1

AN INTRODUCTION TO ENSEMBLE LEARNING IN RESEARCH AND APPLICATIONS

Yi-Tung Chan**, PhD
Department of Electrical Engineering,
R. O. C. Naval Academy, Kaohsiung, Taiwan

ABSTRACT

Ensemble learning (EL) is an important research area for classification in machine learning. Ensemble approaches achieve enhanced performance by combining individual classifiers or training models and utilizing their strengths. In particular, combining the output of several classifiers and models via averaging may reduce the risk of selecting a poorly performing classifier or model and enhance the robustness of the overall accuracy. Ensemble learning algorithms have been successfully applied to decision making, biomedical, financial, remote sensing, genomic, oceanographic, and chemical data analysis and classification problems.

In recent years, with the increasing popularity of central and graphics processing units (CPU/GPU) in computers and other devices and strong

* Corresponding Author's Email: eadown92@gmail.com.

computing capabilities, the use of deep learning and convolutional neural networks has grown rapidly. Among their many applications, computer vision and intelligent video-based systems are particularly important, owing to their effectiveness in providing interpretable applications including intelligent surveillance, face recognition, person re-identification, anomaly detection, image segmentation, video tracking, and intelligent transportation systems.

In this chapter, I introduce the idea that the fundamental aspects of ensemble learning and deep learning may be more beneficial than a single model or classifier. This chapter also explains the benefits of EL in research and applications. In subsequent chapters, worldwide experts and scholars provide insights into EL from various fields. Each chapter opens with an introduction to Mmachine learning and EL techniques and analyzes the applications of EL in different fields, such as signal and image processing, medical care, education, aviation (geology), and agriculture. It is hoped that these applications in various fields can inspire readers to use EL in practice.

Keywords: ensemble learning, meta learner, ensemble system, multiple classifier systems, committee of classifiers, committee-based learning, model ensemble, mixture of experts

1. Introduction

Artificial intelligence (AI) has become a part of our daily lives. As an application of AI, ensemble learning (EL) provides solutions to practical problems in data analysis based on its core philosophy that "two heads are better than one." In AI and machine learning (ML), the "heads" that work together on a solution are diverse "minds" such as learners, agents, and detectors, who classify data and make decisions. According to existing theory, an ensemble system that combines multiple accurate, yet diverse, base classifiers has been proven to enhance the overall classification accuracy. EL is straightforward and practical. Knowledge of EL can help us to obtain relatively high accuracy and to establish a robust ensemble system when facing classification problems. Techniques and concepts related to EL have commonly been used in various international ML and data science competitions to achieve good results. In recent years, EL has been adopted by most of the winners of international data mining competitions, such as competitions hosted by Kaggle and the KDD Cup organized by the Association for Computing Machinery (ACM). For example, a team from National Taiwan University led by professors Chih-Jen Lin, Shou-De Lin, and Hsuan-Tien Lin integrated different models from each of the team's groups that resulted in more precise predictions. This strategy has continuously won them prizes at the ACM KDD Cup over the past six years.

The reader does not need an in depth knowledge of computer science to benefit from this book since EL is a practical and application-oriented technology in ML and most learning algorithms in EL are well developed and verified. The editors of this book and the author of this chapter did not major in computer science. After many years of practice in the industry, I joined a master's degree program in computer science. While studying ML, I was deeply attracted by the theory of ensemble learning. Having worked for many years after graduation, I returned to the university to complete a doctoral degree in computer science and information engineering and then started my career as a university lecturer. As a learner and practitioner of EL for over 12 years, I have benefited from it in terms of resolving practical issues at work and in research. I thus sincerely hope to introduce EL from a practical point of view in order to help you resolve real problems. Research experience or relevant knowledge about computer science and engineering is useful, but a basic understanding of classification algorithms can also facilitate your reading of this book. Ideally, readers will possess a greater degree of understanding of the data and materials used for data analysis than of classification learning algorithms. Invalid datasets can result in defects in the classifiers. As a famous ML saying goes, "Garbage in, garbage out." For the in-depth mathematical foundations and principles of EL, please refer to the recommended reference books and original articles according to your needs. Alternatively, readers may refer directly to the classification algorithms and EL frameworks used by the winners of data science competitions in relevant fields and apply or improve these existing methods to specific problems. For example, readers can use the C4.5 algorithm for decision trees, which is ranked first among the top 10 algorithms in data mining [1] as a base classifier that can be used without the need to implement it in a programming language or in-depth knowledge of C4.5. Ensemble decisions can be obtained after integrating multiple base classifiers through EL. In addition, readers can directly use the Random Forest algorithm, which is a representative method in EL. Both approaches can be used directly to resolve classification problems in practice. The book does not intend to propose a new basic classification algorithm in these chapters but to design integration strategies, learning strategies, and ensemble frameworks that are suitable for resolving practical problems. We hoped that a combination of the research findings of the authors of these different chapters can be used as examples to provide a reference for readers. For details of the algorithms and theory, please refer to the cited publications in the relevant fields; this chapter introduces the essence and concepts of EL.

In this book, readers are encouraged to practice skills as part of the learning process. EL is a practical subject for resolving issues in data analysis. Compared with reading mathematical principles and scientific literature, it is easier to understand the principles of EL by testing them out through practical activities. I recommend that readers use the Waikato Environment for Knowledge Analysis (WEKA), which is an open source machine learning software [2]. It has a friendly GUI interface, rich learning content, and a vibrant discussion community. WEKA can be used directly to conduct data preprocessing and ML modeling for data analysis without programming. If a reader wants to write a program by hand, Python is a highly recommended programming language; it offers rich online resources and multi-level support from the online community, as well as numerous learning and teaching materials. For example, the free 7-day mini-course provided by Machine Learning Mastery (https://machine learningmastery.com/) is suitable for beginners who want to learn ML and program in Python. The best way to learn is to first practice and then to participate in data analysis and competitions in relevant fields. Ultimately, the goal is to apply this knowledge to real issues and the professional field.

Like classification algorithms, in the future, deep learning (DL) may also use EL to build a robust ensemble system for practical applications. Furthermore, the world has entered the era of the Artificial Intelligenceof Things (AIoT) and fifth generation technology (5G). In the future, multi-sensor information fusion will be another important application of EL. Given the strong computing capabilities of Edge AI, DL and AI can be applied to each sensor at the edge to obtain inferences. The inferences of different sensors can then be integrated by EL for ensemble decision making or information fusion. It is believed that this approach might be helpful for applications of AI and ML in real-world environments.

This book consists of two parts. The first part includes Chapters 1, which start from the most basic concepts in AI and ML and introduce the relevant background information with brief descriptions. Chapter 1 introduces data pre-processing, model processing, ensemble learning, test calibration, and the practical deployment of applications in the hope of providing practical applications to meet the needs of readers. The second part comprises Chapters 2 to 8, which present in-depth discussions on the use of EL in different fields by experts and scholars from around the world. I, as the editor of this book, hope that the diversity of applications of EL will further stimulate readers' creative thinking in problem solving, which is the most important purpose of this book and the essence of EL. The authors hope that readers, especially those not majoring in computer science or related subjects, can benefit from the fundamental and practical applications of ensemble learning in this book and can enjoy its beauty.

2. What Is Ensemble Learning?

EL along with DL, which has gained popularity in recent years, are both subsets of ML. AI emerged first followed by ML. DL, which has contributed to the explosive development of AI, falls in the innermost circle. Like DL, EL is a subset of ML in AI. EL, which is a technique that is used to combine multiple classifiers or classification models to obtain more robust results [3], is a research area in ML. It has been successfully applied to data management, data science competitions, and industrial settings. Given its wide range of applications, EL is one of the most important practical techniques in ML and AI. EL can also be called a meta learner, ensemble system, multiple classifier system, committee of classifiers, committee-based learning, model ensemble, or mixture of experts.

Since its emergence in the 1990s, the success of EL has been due to the evolution of classification algorithms in ML and increasing computational capacity thanks to the power of Moore's Law. EL enables data science engineers to combine learning algorithms with weak classifiers to obtain a strong ensemble classifier that enhances practical applications in both hardware and software design. For instance, in financial analysis, every 1% increase in accuracy may greatly reduce investment losses. Even for medical diagnosis, we could save two more lives with every 1% increase in the accuracy of disease classification.

The general machine learning workflow includes two phases: the learning phase and the application phase. First, "machine learning is one capability that is used to acquire knowledge of AI systems by extracting patterns from raw data" [4]. Based on ML algorithms, we can use collected data (or a training dataset) and extracted features to construct training models. In the application phase, the trained model can obtain predictions on new samples (or a testing dataset).

It should be noted that during the learning phase and application phase, the model may require continual iterations and refinements in order to determine the appropriate features, learning algorithms, or parameters. This is done before entering the actual deployment phase in order to establish a problem-specific model. To understand AI and ML, it is first necessary to understand that no universal model exists that can resolve all problems. Instead, we need to first understand the data and then find a suitable learning algorithm. After repeated testing and adjustment, we obtain a model or classifier for resolving a specific problem. This can be explained by the famous "no free lunch" theorem proposed by [5], which states that no single classifier modeling approach is optimal for all pattern recognition tasks.

Extracting representative features from raw data is the key in EL. Nevertheless, extracting high-level, abstract features from raw data is very difficult, and manually designing and labeling features for a complex task requires a lot of manpower, time, energy, and sometimes even decades of effort by the entire research community [4]. Unlike the traditional approach of manually defining features for ML, DL can learn features from the data and express complex representations with other simple features, resolving the core problem of representation learning [4]. Put simply, a convolutional neural network (CNN) can replace the traditional manual feature extraction process in ML with "automatic" image feature extraction. CNNs combine the processes of feature extraction and learning algorithms into one. For more information about DL, the book *Deep Learning* by Goodfellow, Bengio, and Courville [4] is recommended.

According to the type of learning algorithm, machine learning can be divided into three categories: supervised learning, unsupervised learning, and reinforcement learning. The main function of supervised learning is to train a model with labeled data. We can classify different types of animals by training models with classification algorithms using labeled images. In ML, supervised learning is most commonly used to process classification tasks; a large amount of valid labeled data is a prerequisite for building models and classifiers. Typical supervised learning algorithms include support vector machines (SVMs) and decision trees.

Unsupervised learning does not require labeled data, so it saves substantial manual labor. Unlike supervised learning, it classifies unlabeled data based on clusters of similar features. In general, unsupervised learning is less accurate than supervised learning in terms of classification. A representative algorithm of unsupervised learning is k-means clustering.

Reinforcement learning is a type of machine learning in which a computer learns to perform a task through repeated trial-and-error interactions with a dynamic environment. The goal of reinforcement learning is to train players (i.e., ML agents) to complete tasks in a given environment, which includes the surrounding environment of the players and the coach. First, the coach gives a command or prompt, and the player observes and then responds to the coach by taking action. If the actions meet or approach the expectations, the coach may give the player a reward; if not, the player does not receive any rewards or may even receive a penalty. In this example, the goal of reinforcement learning is to adjust or train the players' strategies so that they learn the required actions to maximize the return from the task. After training, the players should be able to take appropriate actions based on their observations of the coach. AlphaGo is one of famous computer programs that utilizes reinforcement learning. Deep learning is another crowning achievement of the extension of the domain of reinforcement learning [4].

Each of the three ML methods mentioned above is associated with numerous techniques and diverse theories, which will not be described in detail in this book. Only the basic concepts of the methods are briefly introduced to facilitate readers' understanding of the subsequent text. For an introduction to ML and its applications, I recommend the Machine Learning Algorithms Mini-Course provided by Machine Learning Mastery [6]. After obtaining a thorough understanding of the basic ML concepts, I recommended that readers put theory into practice for better learning results.

3. How does Ensemble Learning Work?

EL is a technique for reducing generalization error by combining several models. The idea is to train several different models separately and then combine the predictions of all the models on the output for the test examples. Different ensemble methods construct the ensemble of models in different ways. For example, each member of an ensemble can be formed by training a completely different kind of model using a different algorithm or objective function. In general, EL can be divided into the following three categories [20, 24]: bagging (i.e., bootstrap aggregating), boosting, and stacking.

3.1. Bagging

Bagging (short for bootstrap aggregating, also called model averaging) is a method that allows the same kind of model, training algorithm, and objective function to be reused several times [4]. Bagging is a technique for reducing generalization error by combining several models [9]. Different examples are included in each dataset, resulting in differences among the trained models.

Bagging classifiers (or learners) in the base learning pool adopt a parallel structure. Multiple data subsets are randomly built from the training dataset through sampling with replacement, and the subsets are used to train different classifiers. Predictions are then produced for the testing dataset by these classifiers. Finally, an ensemble decision is obtained through a combination method. To combine the predictions obtained from different classifiers, the voting method (for categorical tasks) or averaging (for continuous tasks such as the estimation of probability) can be used. For bagging, a commonly used base learning algorithm is the decision tree. The main representative of bagging is the Random Forest.

3.2. Boosting [10]

Both boosting and bagging randomly create a number of data subsets from the training dataset through sampling with replacement. The difference between boosting and bagging is that boosting further integrates a weighting strategy in the sampling process that assigns higher weights to the incorrectly classified examples to increase the diversity among the different classifiers. Boosting classifiers in the base learning pool have a serial structure: the classification error is measured after each classifier is trained, and the samples that are classified incorrectly by the first classifier receive a larger weight in the subsequent training subsets. The training data subset is updated accordingly for the next classifier. In this way, a new subset is used to train each classifier. Finally, for the combination method, weighted voting (for categorical tasks) or weighted averaging (for continuous values [typically real numbers] tasks) is usually used to combine the predictions obtained from the different classifiers. The main representative of boosting is AdaBoost [11].

Recently, modern boosting methods have been built on AdaBoost, most notably the stochastic Gradient Boosting Machine (GBM) [12]. The name gradient boosting refers to the use of the gradient descent algorithm to minimize the loss when adding new models. The GBM is an implementation of gradient-boosted decision trees designed for speed and performance.

Some EL techniques that have been widely used in recent years are boosting algorithms based on extensions of the GBM, such as categorical boosting (CatBoost) [13], Light Gradient Boosting Machine (LightGBM) [13], and eXtreme Gradient Boosting (XGBoost) [14]. Though CatBoost, LightGBM, and XGBoost are based on the GBM, they address issues associated with the traditional GBM. GBM implementations are slow because of the sequential nature in which each tree is constructed and added to the model. CatBoost, LightGBM, and XGBoost are gradient boosting algorithms based on decision trees. CatBoost is mainly used for categorical tasks; Adaboost, Gradient Boosting Decision Tree (GBDT), and XGBoost can be used for both classification and regression. Adaboost typically uses decision trees as learners to train the classifiers in the base learning pool, while GBDT and XGBoost use classification and regression trees (CART) to construct prediction models [15].

3.3. Stacking

Stacking (short for stacked generalization [16], also known as meta ensembling, meta-learning, stacking meta-learning, or stacked ensembles) isoften based on heterogeneous learning algorithms [7]. Stacking obtains the final ensemble decision by stacking different classifier layers, hence the name. As in bagging, stacked classifiers in the base learning pool have a parallel structure. The difference between stacking and bagging is the algorithm used by the classifiers in the base learning pool. In bagging, each classifier uses the same classification learning algorithm (such as a decision tree), while stacking uses different algorithms to train different classifiers (such as decision trees, SVMs, and neural networks). In other words, each classifier in stacking uses a heterogeneous learning algorithm in contrast to bagging and boosting. In terms of the combination method, stacking also combines the predictions of different classifiers by training classifiers.

Discussion of the mathematical foundations of the above three EL techniques can be found in *Ensemble Methods: Foundations and Algorithms* by Zhi-Hua Zhou [7] and *Boosting: Foundations and Algorithms* co-authored by Robert E. Schapire and Yoav Freund [17].

4. WHERE CAN ENSEMBLE LEARNING BE APPLIED?

In addition to reducing the risk of inference errors, EL has the following five advantages when applied to classification problems [21, 20]:

1. *The ability to handle large training datasets:*

When the number of training datasets is very large, training may be ineffective if only a single classification algorithm is used to train the classification model. To address this issue, the ensemble system can be used to divide a large number of training datasets into multiple training data subsets. The classification algorithm can then be used to train each subset and obtain different classification models. Finally, the results of these classification models are combined to output the overall inference.

2. *The ability to handle imbalanced datasets:*

Imbalanced datasets are commonly encountered in practical applications, and it is challenging to deal with them. When handling imbalanced datasets, the small number of instances in the minority class can result in a lack of data for the minority class during the training of the classification model. In this case, repeated sampling can be used to increase the number of samples from the minority classes to form different training datasets. Classification algorithms can then be used to generate different classification models. This approach can improve the classification performance for minority classes. In addition, the ensemble system can be used to establish top classification models based on different classification algorithms, and the prediction results of these classification models can be combined to improve the classification performance on imbalanced datasets.

3. *The ability to enhance the performance of linear classification models:*

When dealing with a non-linear classification problem, linear classification learning algorithms cannot distinguish the class boundaries. However, if an ensemble system is used in combination with multiple linear classification models, the non-linear classification boundaries can be obtained, and the problem can be distinguished and solved.

4. *Data fusion:*

When facing different datasets and feature sets, an ensemble system can also be used to deal with data fusion. For example, when a bank is evaluating the likelihood of a borrower going into bankruptcy, the training datasets may contain a variety of data, such as personal asset data, records of loans, family data, and work records. Each dataset has unique features. In this case, a single classification model cannot be trained for all of these datasets. However, an ensemble system can train classification models for each dataset separately and then combine the data obtained from the models for the final overall inference. For example, unmanned vehicles are currently equipped with different sensors, such as electro-optical sensors, radar, ultrasound, and LiDAR. It is of great importance to effectively integrate the data collected by these sensors in order to obtain the best information and inference for automatic navigation, collision avoidance, and self-driving.

It's important to note that AI has revolutionized almost every field—from the automotive industry to the electrical industry—continuously affecting the dynamics of society and the lives of consumers in recent years. Among the countless applications of AI, eadge devices and edge computing provides the ideal solution for specific workload acceleration.

The "edge" in edge computing refers to terminal equipment such as mobile communication base stations and environmental monitoring sensors. The opposite of the edge is the cloud server, which is located in the "center." Shifting calculations from the center to the edges eases the load on the cloud server and reduces data processing latency. Therefore, edge AI is being heavily used as an important complement to workstation and cloud computing in applications from traditional monitoring systems to the unmanned vehicles that are currently being developed.

In the future, EL techniques can be adopted to further combine heterogeneous information from multiple edge AI devices. The different devices can complement each other and can be used to form an ensemble system for all kinds of data analysis and industrial applications. It is believed that the way to best integrate edge computing devices with EL techniques will become an important research direction for AI in the future.

5. Ensemble Learning and Deep Learning:

With the emergence of big data, increasing computing power, and the development of the graphics processing unit (GPU), AI and DL have seen great success in recent years and have become important research areas. It is worth noting that EL also plays an important role in this success. In particular, the architecture and concepts of EL that have been designed to avoid overfitting have been successfully applied to DL. For instance, Schwenk and Bengio et al. [24] have applied boosting to build ensembles of neural networks by incrementally adding neural networks to the ensemble. Boosting has also been applied by interpreting an individual neural network as an ensemble by incrementally adding hidden units to the network [4].

CONCLUSION

As Dr. John L. Hennessy wrote in *Leading Matters: Lessons from My Journey*: “making decisions with the power of the people is one of the reasons for human progress.” The same is true for EL. EL is a broad theory and is not limited to fixed models or algorithms. It tends to focus on an idea, which is "two heads are better than one," and it is not restricted to existing EL algorithms (such as the Random Forest). The key point is integrating models on the basis of EL algorithms to achieve better results with the ensemble system. This chapter provides references and recommended reading for each technique to satisfy the curiosity of some readers with regard to mathematical theories and algorithms so that they can acquire further knowledge and answer their questions. In particular, the book Introduction to Data Mining written by Professor Zhi-Hua Zhou from Nanjing University in 2014 is one of the classics in EL studies. This book deserves to be known as a classic as it provides a detailed introduction of the history of EL, guidance on mathematical theories, and analysis of classical algorithms. Finally, the hope is that readers can be made aware, through practical use of EL, that they can build a robust ensemble system and solve problems in their areas without having to learn the absolute details of specific ML algorithms and mathematics behind the algorithms.

This book dedicates substantial space to describing the application of EL in various fields. In practice, the actual environment and data resources in scientific research and applications may differ from those in traditional textbooks, so learners may not obtain the expected results. To this end, this book introduces EL with examples of real-life applications to facilitate the readers' understanding and meet their practical needs.

I strongly encourage readers to practice. If you do not have the appropriate materials on hand, datasets can be downloaded from the UCI Machine Learning Repository [19] for practice. It is also beneficial to participate in data science competitions in various fields. In addition to the previously mentioned KDD-Cup [20] and Kaggle [21], DrivenData [22] also hosts data science competitions in various fields, such as international development, medical care, education, research and protection, and public services. InnoCentive [23] provides data analysis in the field of life sciences, including data analysis competitions related to the global COVID-19 pandemic in 2020. Of course, the problems in a competition are not necessarily the same as the real problems, but these competitions allow you to put the knowledge you have learned into practice. It is a good opportunity for not only learning AI, ML, and EL but also for accessing the best methods and data resources available to date.

Chapters 2 to 8 present works of scholars from various fields. In addition to providing readers with references to examples of practical applications in various fields, we hope this book can further inspire readers to apply EL and related concepts to their field in order to resolve actual issues in data analysis. This book extensively introduces and discusses the application of EL in various fields and the current and future research directions of its novel applications. It also reviews some of the more popular areas in which EL has received widespread attention in recent years in the ML and AI communities.

REFERENCES

[1] Wu, X. et al. 2008. "Top 10 algorithms in data mining," *Knowl. Inf. Syst., vol. 14*, no. 1, pp. 1–37.

[2] "*Weka 3—Data Mining with Open Source Machine Learning Software in Java*." Last modified 04 April 2020. https://www.cs.waikato.ac.nz/ml/weka/.

[3] Dietterich, T. G. 2000. "Ensemble methods in machine learning," in *International workshop on multiple classifier systems,* pp. 1–15.

[4] Goodfellow, I., Y. Bengio, and A. Courville. 2016. *Deep Learning*. MIT Press.

[5] Wolpert, D. H. 2002. "The Supervised Learning No-Free-Lunch Theorems." In *Soft Computing and Industry: Recent Applications*, edited by R. Roy, M. Köppen, S. Ovaska, T. Furuhashi, and F. Hoffmann, 25-42. London: Springer London, 2002.

[6] "*Machine Learning Algorithms Mini-Course*." Last modified 22 May 2020. https://machinelearningmastery.com/machine-learning-algorithms-mini-course/.

[7] Li, Z. and X. Yan. 2019. "Ensemble model of wastewater treatment plant based on rich diversity of principal component determining by genetic algorithm for status monitoring," *Control Eng. Pract.*, vol. 88, pp. 38–51.

[8] Galar, M., A. Fernandez, E. Barrenechea, H. Bustince, and F. Herrera. 2012. "A Review on Ensembles for the Class Imbalance Problem:

Bagging-, Boosting-, and Hybrid-Based Approaches." *IEEE Trans. Syst. Man, Cybern. Part C* (Applications Rev.) 42:4:463–484.

[9] Breiman, L. 1996. "Bagging predictors." *Mach. Learn*. 24:2:123–140.

[10] Freund, Y. and Schapire E. Robert 1996. "Experiments with a new boosting algorithm." In *Machine Learning: Proceedings of Thirteenth International Conference*: 148–156.

[11] Freund, Y. and R. E. Schapire. 1995. "A decision-theoretic generalization of on-line learning and an application to boosting," in *European conference on computational learning theory*, pp. 23–37.

[12] Breiman, L., J. Friedman, C. J. Stone, and R. A. Olshen. 1984. *Classification and regression trees*. CRC Press.

[13] Prokhorenkova, L., G. Gusev, A. Vorobev, A. V. Dorogush, and A. Gulin. 2018. "CatBoost: unbiased boosting with categorical features." In *Advances in Neural Information Processing Systems* 31, edited by S. Bengio, H. Wallach, H. Larochelle, K. Grauman, N. Cesa-Bianchi, and R. Garnett, 6638–6648. Curran Associates, Inc.

[14] Chen, T. and C. Guestrin. 2016. "XGBoost." *Proceedings of the 22nd ACM SIGKDD International Conference on Knowledge Discovery and Data Mining* 13-17:785–794.

[15] Breiman, L., J. Friedman, C. J. Stone, and R. A. Olshen. 1984. *Classification and regression trees*. CRC Press.

[16] Wolpert, D. H. 1992. "Stacked generalization." *Neural Networks* 5:2:241–259.

[17] Schapire E. Robert and Freund, Y. 2013. "Boosting: Foundations and Algorithms." *Kybernetes* 42:1:164–166.

[18] Polikar, R. 2006. "Ensemble based systems in decision making." *IEEE Circuits Syst. Mag*. 6:3:21–45.

[19] "*UCI Machine Learning Repository*." Last modified 09 April 2020. https://archive.ics.uci.edu/ml/index.php.

[20] "SIGKDD–KDD Cup." Last modified 22 May 2020. https://www.kdd.org/kdd-cup.

[21] "*Kaggle: Your Machine Learning and Data Science Community*." Last modified 22 May 2020. https://www.kaggle.com/.

[22] "*Competitions*." Last modified 23 May 2020. https://www.drivendata.org/competitions/.

[23] "*InnoCentive | Open Innovation & Crowdsourcing Platform*." Last modified 23 May 2020. https://www.innocentive.com/.

[24] Schwenk, H. and Y. Bengio. 1998. "Training methods for adaptive boosting of neural networks." *Adv. Neural Inf. Process. Syst*. 647–650.

In: An Introduction to Approaches
Editor: Yi-Tung Chan
ISBN: 978-1-53618-680-2

Chapter 2

ENSEMBLE DENOISING AUTOENCODERS: ENSEMBLE LEARNING FOR A NOISE REDUCTION

Kazuki Sakai[1,*] ***Yuto Omae***[2,†] ***and Hirotaka Takahashi***[3,4,‡]
[1]Department of Electronic Control Engineering,
National Institute of Technology, Nagaoka College,
Nagaoka, Niigata, Japan
[2]Department of Industrial Engineering and Management,
College of Industrial Technology, Nihon University,
Narashino, Chiba, Japan
[3]Department of Information and Management Systems Engineering,
Nagaoka University of Technology, Nagaoka, Niigata, Japan
[4]Research Center for Space Science, Advanced Research Laboratories,
Tokyo City University, Setagaya, Tokyo, Japan

Abstract

In the domain of engineering, especially in sensing or control tasks, it is necessary to analyze time-series data that contain noise originating from various sources, such as disturbances and measurement errors. Therefore, techniques that precisely reduce noise in observed data, or

*Corresponding Author's Email: k-sakai@nagaoka-ct.ac.jp.
†Corresponding Author's Email: oomae.yuuto@nihon–u.ac.jp.
‡Corresponding Author's Email: hirotaka@tcu.ac.jp.

high-quality noise-reduction techniques, can improve both sensing and control accuracies. Noise reduction is one of the tasks for which machine-learning techniques are assumed to be effective. However, ensemble learning has fewer applications in noise-reduction problems as compared with its applications in classification or regression problems, although ensemble learning has been used widely to enhance the generalization ability on both classification and regression problems. In this chapter, we introduce an ensemble-learning-based noise-reduction method. One of the representative methods of machine-learning-based noise reduction is the denoising autoencoder (DAE). The DAE is an extension of an autoencoder, which is a kind of a neural network used for dimensional reduction. The DAE is constructed to output an original signal when a superposition of the signal and noise is input.

Keywords: Denoising Autoencoder (DAE), Neural Network (NN), Noise Reduction

INTRODUCTION

In the domain of engineering, especially in sensing or control tasks, it is necessary to analyze time-series data that contains noise originating from various sources, such as disturbances and measurement errors. For example, in the research of gravitational waves, the gravitational wave signals are so weak that the noise from the detector itself dominates the observed data; therefore, the presence of noise reduces the estimation accuracy of gravitational-wave-source parameters. Digital filters are the conventional means to manipulate such noise. Using band-pass filters and/or band-stop filters, we can attenuate specific frequency bands in which noise seems to dominate. However, the optimal design of the filters varies depending on data, hence, the filter design also depends on the experience of a designer.

In 2008, Vincent *et al.* proposed the denoising autoencoder (DAE) as a new approach to achieving noise reduction by using the neural network (NN) [1], which is one of the machine-learning methods. Its structure comprises the repetition of a combination of matrix operation and non-linear transformation. In addition, it can express arbitrary input–output relations when the elements of matrices are appropriately determined. Recently, various NN-based research works, such as deep neural networks (DNNs) and utilization of graphical processing units to accomplish rapid calculation have made satisfactory progress. The NN, especially the DNN, is achieving popularity nowadays because it can achieve high accuracy and generalization without the manual extraction of feature values.

However, ensemble learning is another way to improve the generalization ability of machine-learning methods. In the ensemble learning, the system contains multiple learners, and its output is a statistical value of their outputs. The essential point is that the learners are trained independently with each other; for example, they use different seed values for their initialization, or they use different datasets for their training. It decreases the influence of over learning for each learner. For instance, the random forest and AdaBoost, both of which have been used widely, are the applications of ensemble learning. In this chapter, we introduce an ensemble-learning-based noise-reduction method. We observed that the denoising performance could be enhanced by lining up multiple DAEs. We named this arrangement of multiple DAEs as the ensemble DAE (E-DAE). Furthermore, we also report the quantitative evaluation of the E-DAE.

In the second section, we briefly describe the NN and its parameter-optimization procedure. The definitions of the autoencoder, DAE, and E-DAE are described in the third section. In the fourth section, as the evaluation of the model, we report the simulation of noise reduction by the E-DAE with a sine-Gaussian signal and stationary Gaussian noise. The last section comprises the summary.

REGRESSION USING NN

Basics of NNs

In 1986, Rumelhart et al. proposed the back-propagation learning algorithm of the NN [2]. This invention rapidly made the NN get famous, and the NN has been applied in various fields since that. There have been some review papers summarizing the applications in each field. For example, Adeli summarized the applications in the field of civil engineering [3], Christodoulou summarized the applications in the electromagnetics' field [4], Yagawa summarized the applications in the computational mechanics' field [5], and Hu summarized the applications in the signal processing field [6]. As this fact that there have been a lot of applications of the NN indicates, the range of the application for the NN is very wide.

In addition to the spread of the applications, the theoretical framework of the NN also has been progressing. The most impactful development is the DNN. In

2012, a team of professor Hinton, who is the co-author of the paper of the back-propagation, won the highest score in the contest "imagenet large scale visual recognition (ILSVR)" by using a method named Supervision, which is based on the DNN [7]. After that, the DNN took the place of the other conventional methods like the support vector machine as the major method in the contest. Even now, the NN technique continues to develop.

As for our target, noise reduction, some applications of the NN have been reported. For example, Lu et al. applied the NN to a noise reduction on the human voice data [8] [9], and Xiong et al. worked on a noise reduction of measured data of electrocardiogram [10]. The effectiveness of the NN in the field of noise reduction also has been confirmed.

The DAE is one of NN-based regression models. Regression is a machine-learning task aiming to predict some values. In this subsection, we describe the basics of NNs, by focusing on a three-layered NN; the three layers are namely, the input layer, hidden layer, and output layer. Let n_1, n_2, and n_3 denote the numbers of units in the input, hidden, and output layers, respectively. We denote the units in the layers as vectors $\boldsymbol{x}_1$, $\boldsymbol{x}_2$, and $\boldsymbol{x}_3$. We have:

$$\boldsymbol{x}_1 = \left[x_{1,1} \ \cdots \ x_{1,n_1}\right]^\mathsf{T}, \tag{1}$$

$$\boldsymbol{x}_2 = \left[x_{2,1} \ \cdots \ x_{2,n_2}\right]^\mathsf{T}, \tag{2}$$

$$\boldsymbol{x}_3 = \left[x_{3,1} \ \cdots \ x_{3,n_3}\right]^\mathsf{T}. \tag{3}$$

A NN that comprises multiple hidden layers is called the DNN. Although the DNN can express more complicated input–output relationships, it requires considerable amount of training data.

The numbers of units in the input and output layers are determined according to the property of the considered problem. For example, if we consider an estimation problem to predict a discomfort index on the basis of the temperature and humidity in a room, the output value is the discomfort index and the input values are the temperature and humidity. Therefore, the input layer has two units corresponding to the temperature and humidity, and the output layer has only one unit corresponding to the discomfort index. The number of units in the hidden layer, n_2, is one of the hyperparameters of the model.

Next, we describe the calculation algorithm of the estimated value. First, data are given to the units in the input layer, $\boldsymbol{x}_1$. Subsequently, the values of the units in the hidden layer, $\boldsymbol{x}_2$, are calculated using $\boldsymbol{x}_1$. Finally, $\boldsymbol{x}_3$, which are the units in the output layer, are calculated using $\boldsymbol{x}_2$. Because the calculations propagate from the input layer to the output layer, as mentioned in the aforementioned sentences, this algorithm is called forward propagation.

Specifically, the j-th unit in the hidden layer, $x_{2,j}$ and the j-th unit in the output layer, $x_{3,j}$, are calculated as follows:

$$x_{2,j} = f\left(\sum_{k=1}^{n_1} w_{jk}x_{1,k} + b_j\right), \tag{4}$$

$$x_{3,j} = f\left(\sum_{k=1}^{n_2} v_{jk}x_{2,k} + c_j\right), \tag{5}$$

where w_{jk} and v_{jk} denote the (j, k) elements of matrices $\boldsymbol{W}$ and $\boldsymbol{V}$, respectively, and b_j and c_j denote the j-th element of vectors $\boldsymbol{b}$ and $\boldsymbol{c}$, respectively. These are the model parameters of the NN. The elements of matrices $\boldsymbol{W}$ and $\boldsymbol{V}$ are called weight parameters, and the elements of vectors $\boldsymbol{b}$ and $\boldsymbol{c}$ are called bias parameters. In addition, $f(z)$ denotes a non-linear real function called the activation function. We introduce vectors $\boldsymbol{h}_1$ and $\boldsymbol{h}_2$, and define their elements as follows:

$$h_{1,j} = \sum_{k=1}^{n_1} w_{jk}x_{1,k} + b_j, \tag{6}$$

$$h_{2,j} = \sum_{k=1}^{n_2} v_{jk}x_{2,k} + c_j. \tag{7}$$

Eqs. (4) and (5) can be expressed as follows:

$$\boldsymbol{x}_2 = f(\boldsymbol{h}_1), \qquad \boldsymbol{h}_1 = \boldsymbol{W}\boldsymbol{x}_1 + \boldsymbol{b}, \tag{8}$$

$$\boldsymbol{x}_3 = f(\boldsymbol{h}_2), \qquad \boldsymbol{h}_2 = \boldsymbol{V}\boldsymbol{x}_2 + \boldsymbol{c}. \tag{9}$$

It means that the outputs of an NN depend on the model parameters. Upon obtaining appropriate model parameters, the NN achieves accurate estimation.

In this study, the following sigmoid function:

$$f(z) = \frac{1}{1 + \mathrm{e}^{-z}}, \tag{10}$$

is used as the activation function. It is one of the most popular activation functions. As for the DNN, the rectified linear unit is more preferred to the sigmoid function [11]. Because the range of the sigmoid function is $(0, 1)$, we must normalize data when using the sigmoid function as the activation function of the output layer.

Cost Function

To obtain appropriate model parameters, the parameters are iteratively updated to improve the performance of the NN on the basis of a performance-representing index value, which is called the cost function. This iterative updation of the parameters is called the training process, and the data used in the training process are called the training data. In a regression problem, training data comprises a pair of the input data $\boldsymbol{x}$ and the correct values $\boldsymbol{y}$ corresponding to $\boldsymbol{x}$. The numbers of units, n_1 and n_3, are set to the dimensions of $\boldsymbol{x}$ and $\boldsymbol{y}$, respectively. Regarding the previous example, $\boldsymbol{x}$ comprises temperature and humidity, and $\boldsymbol{y}$ comprises the correct discomfort index of the atmosphere. For an effective training process, an enormous amount of training data is required. We denote the dataset of the collected training data as D_{T} and call it the training dataset. We have the following:

$$D_{\mathrm{T}} = \left\{ (\boldsymbol{x}^{(i)}, \boldsymbol{y}^{(i)}) | i = 1, \ldots, N_{\mathrm{T}} \right\}, \tag{11}$$

where $(\boldsymbol{x}^{(i)}, \boldsymbol{y}^{(i)})$ denotes a pair of the i-th data and the correct value thereof, and N_{T} denotes the amount of training data.

Basically, the mean squared error is used as the cost function of the regression problem. Let $\boldsymbol{x}_3^{(i)}$ denote the output when $\boldsymbol{x}^{(i)}$ is input to the NN. The cost function L is then defined as follows:

$$L = \frac{1}{N_{\mathrm{T}}} \sum_{i=1}^{N_{\mathrm{T}}} L^{(i)}, \tag{12}$$

$$L^{(i)} = \frac{1}{n_3} \left\| \boldsymbol{x}_3^{(i)} - \boldsymbol{y}^{(i)} \right\|^2 = \frac{1}{n_3} \sum_{m=1}^{n_3} \left(x_{3,m}^{(i)} - y_m^{(i)} \right)^2. \tag{13}$$

The lower the value of L, the more accurate is the estimation achieved by the NN.

Parameter Learning

In this subsection, we describe how to update the model parameters in order to decrease the cost function L. The gradient descent is one of the widely used parameter-learning methods [12]. Let $\theta^{(t)}$ denote a value of parameter θ at the

t-th time in the learning process. The parameter is then updated as follows:

$$\theta^{(t+1)} = \theta^{(t)} - \gamma \frac{\partial L(\theta = \theta^{(t)})}{\partial \theta}, \tag{14}$$

where γ denotes a hyperparameter called learning rate, which controls the speed and stability of the learning process.

After the parameter-learning process iterates sufficient number of times, we obtain the parameter values that minimize the cost function. In the 3-layered NN, θ can be any of the elements of $\boldsymbol{W}$, $\boldsymbol{V}$, $\boldsymbol{b}$, and $\boldsymbol{c}$. The number of parameters to be updated is:

$$n_1 \times n_2 + n_2 \times n_3 + n_2 + n_3. \tag{15}$$

To execute the gradient descent, we calculate $\partial L / \partial \theta$ for each of them. In the remaining subsection, we explain an efficient way to calculate $\partial L / \partial \theta$.

First, we here list some partial-differentiation notations used in this discussion as follows:

$$\frac{\partial L}{\partial L^{(i)}} = \frac{1}{N_{\mathrm{T}}}, \qquad \frac{\partial L^{(i)}}{\partial x^{(i)}_{3,m}} = \frac{2}{n_3}\left(x^{(i)}_{3,m} - y^{(i)}_m\right), \tag{16}$$

$$\frac{\partial x^{(i)}_{3,m}}{\partial h^{(i)}_{2,j}} = f'\left(h^{(i)}_{2,m}\right)\delta_{mj}, \qquad \frac{\partial x^{(i)}_{2,m}}{\partial h^{(i)}_{1,j}} = f'\left(h^{(i)}_{1,m}\right)\delta_{mj}, \tag{17}$$

$$\frac{\partial h^{(i)}_{2,m}}{\partial x^{(i)}_{2,j}} = v_{mj}, \qquad \frac{\partial h^{(i)}_{1,m}}{\partial x^{(i)}_{1,j}} = w_{mj}, \tag{18}$$

$$\frac{\partial h^{(i)}_{2,m}}{\partial v_{jk}} = x^{(i)}_{2,k}\delta_{mj}, \qquad \frac{\partial h^{(i)}_{1,m}}{\partial w_{jk}} = x^{(i)}_{1,k}\delta_{mj}, \tag{19}$$

$$\frac{\partial h^{(i)}_{2,m}}{\partial c_j} = \delta_{mj}, \qquad \frac{\partial h^{(i)}_{1,m}}{\partial b_j} = \delta_{mj}, \tag{20}$$

where δ_{mj} denotes the Kronecker delta. It is confirmed that $\partial L / \partial \theta$ denotes the mean value of $\partial L^{(i)} / \partial \theta$ for all the training data. Therefore, we only have to establish a way to calculate $\partial L^{(i)} / \partial \theta$.

We introduce new variables $\Delta^{(i)}_{1,j}$ and $\Delta^{(i)}_{2,j}$ defined as follows:

$$\Delta^{(i)}_{1,j} = \frac{\partial L^{(i)}}{\partial h^{(i)}_{1,j}}, \qquad \Delta^{(i)}_{2,j} = \frac{\partial L^{(i)}}{\partial h^{(i)}_{2,j}}, \tag{21}$$

and vectors $\boldsymbol{\Delta}_1^{(i)}$ and $\boldsymbol{\Delta}_2^{(i)}$ are defined as follows:

$$\boldsymbol{\Delta}_1^{(i)} = \left[\Delta_{1,1}^{(i)}, \ldots, \Delta_{1,n_2}^{(i)}\right]^{\mathsf{T}}, \qquad \boldsymbol{\Delta}_2^{(i)} = \left[\Delta_{2,1}^{(i)}, \ldots, \Delta_{2,n_3}^{(i)}\right]^{\mathsf{T}}. \tag{22}$$

By using these, $\partial L^{(i)}/\partial w_{jk}$ can be calculated as follows:

$$\begin{aligned}\frac{\partial L^{(i)}}{\partial w_{jk}} &= \sum_{m=1}^{n_2} \frac{\partial L^{(i)}}{\partial h_{1,m}^{(i)}} \frac{\partial h_{1,m}^{(i)}}{\partial w_{jk}} \\ &= \sum_{m=1}^{n_2} \Delta_{1,m}^{(i)} x_{1,k}^{(i)} \delta_{mj} \\ &= \Delta_{1,j}^{(i)} x_{1,k}^{(i)}.\end{aligned} \tag{23}$$

Similarly, the following equations hold:

$$\frac{\partial L^{(i)}}{\partial v_{jk}} = \Delta_{2,j}^{(i)} x_{2,k}^{(i)}, \tag{24}$$

$$\frac{\partial L^{(i)}}{\partial b_j} = \Delta_{1,j}^{(i)}, \tag{25}$$

$$\frac{\partial L^{(i)}}{\partial c_j} = \Delta_{2,j}^{(i)}. \tag{26}$$

These indicate that $\partial L^{(i)}/\partial \theta$ can be calculated easily when $\boldsymbol{\Delta}_1^{(i)}$ and $\boldsymbol{\Delta}_2^{(i)}$ are given. Furthermore, using the definition of $\Delta_{2,j}^{(i)}$, it can be easily calculated as follows:

$$\begin{aligned}\Delta_{2,j}^{(i)} &= \frac{\partial L^{(i)}}{\partial h_{2,j}^{(i)}} \\ &= \sum_{m=1}^{n_3} \frac{\partial L^{(i)}}{\partial x_{3,m}^{(i)}} \frac{\partial x_{3,m}^{(i)}}{\partial h_{2,j}^{(i)}} \\ &= \sum_{m=1}^{n_3} \frac{2}{n_3} \left(x_{3,m}^{(i)} - y_m^{(i)}\right) f'\left(h_{2,m}^{(i)}\right) \delta_{mj} \\ &= \frac{2}{n_3} \left(x_{3,j}^{(i)} - y_j^{(i)}\right) f'\left(h_{2,j}^{(i)}\right).\end{aligned} \tag{27}$$

As for $\Delta_{1,j}^{(i)}$, we have the following:

$$\begin{aligned}
\Delta_{1,j}^{(i)} &= \frac{\partial L^{(i)}}{\partial h_{1,j}^{(i)}} \\
&= \sum_{m=1}^{n_3} \frac{\partial L^{(i)}}{\partial h_{2,m}^{(i)}} \sum_{l=1}^{n_2} \frac{\partial h_{2,m}^{(i)}}{\partial x_{2,l}^{(i)}} \frac{\partial x_{2,l}^{(i)}}{\partial h_{1,j}^{(i)}} \\
&= \sum_{m=1}^{n_3} \Delta_{2,m}^{(i)} \sum_{l=1}^{n_2} v_{ml} f'\left(h_{1,l}^{(i)}\right) \delta_{lj} \\
&= f'\left(h_{1,j}^{(i)}\right) \sum_{m=1}^{n_3} \Delta_{2,m}^{(i)} v_{mj}. \qquad (28)
\end{aligned}$$

It means that the following relationship holds:

$$\boldsymbol{\Delta}_1^{(i)} = \boldsymbol{V}^\top \boldsymbol{\Delta}_2^{(i)} \odot f'\left(\boldsymbol{h}_1^{(i)}\right), \qquad (29)$$

where $\boldsymbol{a} \odot \boldsymbol{b}$ denotes the Hadamard product between vectors $\boldsymbol{a}$ and $\boldsymbol{b}$, and it is defined as follows:

$$[a_1 \cdots a_n]^\top \odot [b_1 \cdots b_n]^\top = [a_1 b_1 \cdots a_n b_n]^\top. \qquad (30)$$

The term $\Delta_1^{(i)}$ can be calculated using the matrix operation called backpropagation.

In this study, we use the sigmoid function as the activation f. The derivative $f'(z)$ is expressed as follows:

$$f'(z) = f(z)(1 - f(z)). \qquad (31)$$

In this case, we first calculate $\Delta_{2,j}^{(i)}$ as follows:

$$\begin{aligned}
\Delta_{2,j}^{(i)} &= \frac{2}{n_3}\left(x_{3,j}^{(i)} - y_j^{(i)}\right) f\left(h_{2,j}^{(i)}\right)\left(1 - f\left(h_{2,j}^{(i)}\right)\right) \\
&= \frac{2}{n_3}\left(x_{3,j}^{(i)} - y_j^{(i)}\right) x_{3,j}^{(i)}\left(1 - x_{2,j}^{(i)}\right), \qquad (32)
\end{aligned}$$

following which the others are calculated using the backpropagation method.

Applied Techniques for Parameter Learning

In the previous subsection, we explained the fundamental training process of the NN. However, in this subsection, we introduce some applied techniques for the training process.

Simulated Annealing

Simulated annealing is an algorithm for approximating the optimal value of a given function [13]. In an NN, simulated annealing can be used as a method to update the model parameters. The parameter update formula is replaced with the following:

$$\theta^{(t+1)} = \theta^{(t)} - \gamma \frac{\partial L(\theta = \theta^{(t)})}{\partial \theta} + \varepsilon^{(t)}, \tag{33}$$

where

$$\varepsilon^{(t)} = r^{(t)} \sqrt{T^{(t)}}, \tag{34}$$

$$T^{(t)} = \frac{T_0}{1+t}, \tag{35}$$

$$r^{(t)} \sim \mathrm{N}(0, 1). \tag{36}$$

$T^{(t)}$ is called the temperature parameter at the t-th time. In addition, T_0, which denotes the initial value of $T^{(t)}$, is a hyperparameter.

In the gradient descent, basically, the training process converges to the local minimum nearest to the initial point. Substituting the value of random variable $\varepsilon^{(t)}$ in Eq. (14) allows to escape from a local minimum. However, simultaneously, the random variable worsens L. Multiplying the square root of the temperature parameter $T^{(t)}$, which is monotonically decreasing with t, the effect of the random variable gets smaller after updating sufficient number of times.

Momentum

The momentum technique expedites the training process [14]. The parameter-update formula by the momentum is expressed as follows:

$$\theta^{(t+1)} = \theta^{(t)} + v^{(t+1)}, \tag{37}$$

$$v^{(t+1)} = \eta v^{(t)} - \gamma \frac{\partial L(\theta = \theta^{(t)})}{\partial \theta}, \tag{38}$$

where $v^{(t)}$ denotes the momentum term at the t-th time. Because the momentum term has a recursive structure, the update direction in the parameter space is determined on the basis of not only the current gradient but also the past results. The term η denotes a hyperparameter that assumes a real value ranging from 0 to 1. If we set $\eta = 0$, the update formula is identical to that of the gradient descent.

Batch, Mini-Batch Learning and SGD

As we explained in the previous section, parameter learning starts from calculation of the derivative $\partial L/\partial\theta$ for the training dataset D_{T}. Batch learning [15] is a method using all data in D_{T} for one step of the parameter learning. Eq. (12) implicitly presupposes batch learning.

Generally, an optimization problem aims to estimate the global minimum of an objective function in the parameter space. However, in machine learning, the final aim is not to estimate the model parameters that globally minimize the cost function L for a given training dataset D_{T}. When the training dataset contains abnormal data, the generalization ability of the NN drops if it attempts to fit itself to the abnormal data; it is a well-known phenomenon called over-fitting or over-learning [16]. The batch learning cannot avoid the effect of the abnormal data. The situation that the training dataset contains abnormal data is not unusual; it is caused by the fuzziness of human judgment, malfunction of measuring instruments, measurement errors, erroneous input, among others. An example situation is that there exists an abnormal person who likes extremely hot and humid circumstances, and the problem is to estimate discomfort index by using the temperature and humidity in a room. In this case, the NN may be trained to say "it is comfortable" when high temperature and humidity are input.

One of the methods to prevent overfitting is mini-batch learning [17]. As the preparation of the mini-batch learning, the sub-training datasets, $D_{\mathrm{mb}}^{(n)}$ $(n = 1, \ldots, N_{\mathrm{mb}})$, which are called mini-batches, are constructed via random sampling from the training dataset D_{T}. Hence, the following relationship holds:

$$D_{\mathrm{T}} = \bigcup_{n=1}^{N_{\mathrm{mb}}} D_{\mathrm{mb}}^{(n)}, \tag{39}$$

where N_{mb} denotes the number of sub-training datasets. The data quantity of each mini-batch is called batch-size; it is a hyperparameter. Let the batch-size be denoted by $N_{\rm bs}$. We have the following:

$$N_{\rm mb} = \left\lfloor \frac{N_{\rm T}}{N_{\rm bs}} \right\rfloor, \quad (40)$$

where $\lfloor x \rfloor$ denotes the floor function of x, which returns the maximum integer smaller than x. In the mini-batch learning, parameter learning is conducted with each mini-batch. The model parameters are updated on the basis of the value of the cost function calculated with the first mini-batch, following which they are updated again on the basis of the value with the next mini-batch. The update is repeated until all the mini-batches are used. Upon using all the mini-batches, they are reconstructed via random sampling. It means that the model parameters are updated $N_{\rm mb}$ times for one realization of mini-batches. We call the updates by one realization of mini-batches as "1 epoch."

Stochastic gradient descent (SGD) [17] is another popular parameter learning algorithm. In the parameter learning by the SGD, the model parameters are updated after each input. The SGD can be described as a special case of the mini-batch learning with the batch-size, $N_{\rm bs}$, being 1. Therefore, the model parameters are updated $N_{\rm T}$ times per epoch.

Both the mini-batch learning and SGD prevent overfitting because a different dataset is used for each update of the model parameters. Notably, the simulated annealing and momentum techniques are the modifications of the parameter-update formula; we can use both the types of modifications. For example, a combination of the momentum technique and SGD is called momentum SGD [18, 19].

Ensemble Learning

Ensemble learning is a method to determine the final output by using statistical processing, such as majority voting or averaging, of outputs from multiple NNs, each of which is called a weak learner [20]. There are some approaches to make weak learners and to determine the final output, such as bagging, boosting, and stacking.

In the case of bagging [21], or bootstrap aggregating, each weak learner is developed independently with each other, and the final output is calculated by taking the mean value of the outputs of the weak learners.

As for boosting [21], weak learners are developed one after another. The first weak learner is trained with a sub-dataset obtained from a training dataset by the bootstrap sampling (explained later). Then we record which data the weak learner failed to predict accurately. The probability for the data in which the first weak learner failed to output the correct value to be sampled into the second sub-dataset is set to higher than the other data. The second weak learner is trained with this second sub-dataset. As a result, the second weak learner may compensate for the weak point of the first weak learner. By repeating this process, data relatively difficult to be correctly predicted are likely to appear in the sub-datasets. As described above, weak learners in the boosting compensate the weak point of the other weak learners, whereas weak learners in the bagging are independent with each other.

The stacking technique uses weak learners and an upper-layer learner to predict the final output from the outputs of the weak learners [15]. In other words, the upper-layer learner uses the outputs of the weak learners as its feature values. Unlike the bagging and boosting, weak learners do not have to be based on the identical model in the stacking. We can combine the NN and the SVM, for example.

We explain a specific process of ensemble learning. First, N_{en} sub-training datasets are created via random sampling from the training dataset D_{T}. We have the following:

$$D_{\mathrm{en}} = \left\{ D_{\mathrm{en}}^{(1)}, \ldots, D_{\mathrm{en}}^{(N_{\mathrm{en}})} \right\}. \tag{41}$$

Unlike in the case of mini-batch learning, the same sample can be in two or more training datasets. In the domain of machine learning, this process is called bootstrap sampling [15].

In the ensemble learning explained above, N_{en} NNs are prepared, each of which is trained using a different sub-training dataset. Consequently, N_{en} outputs are obtained from the NNs by feeding one input. Subsequently, the final output is calculated via specific statistical processing of these outputs. For effective ensemble learning, the correlations between weak learners should be low. Actually, it is a desirable situation that each learner falls into overfitting to its dataset. Furthermore, the number of ensembles, N_{en}, is a hyperparameter.

In addition, we can define a dataset, $O^{(n)}$, satisfying the following condition:

$$D_{\mathrm{T}} = D_{\mathrm{en}}^{(n)} \cup O^{(n)}, \tag{42}$$

where $O^{(n)}$ denotes the n-th out-of-bag dataset. Because $O^{(n)}$ is not used in the parameter learning of the n-th weak learner, the former can be used as a validation dataset for searching better hyperparameters of the learner. For example, the number of units in the hidden layer, n_2, can be selected using the following procedure. We construct NNs with various values of n_2, train them using D_{en}, and finally evaluate the error rate by using the out-of-bag datasets. The error rate is called the out-of-bag error [15]; it is known that the out-of-bag error is highly related to the generalization error of the entire system. The number of units in the hidden layer resulting in the minimum out-of-bag error is the possible best value of n_2.

Although introducing ensemble learning often provides high-quality estimation, it takes longer time than that taken by a single NN for both parameter-learning and estimation processes.

ENSEMBLE DAE

In the section named "REGRESSION USING THE NN", we explained the NN for solving the regression problem. In this section, we first describe autoencoder and DAE on the basis of the relationship of the Reg-NN, following which we introduce the definition of our proposed model, i.e., ensemble denoising autoencoder (E-DAE).

Autoencoder

Autoencoder, which is a kind of Reg-NN, is typically used for dimension reduction or pre-training a DNN. Because an autoencoder belongs to the class of un-supervised learning, its training dataset can be expressed as follows:

$$D_{\text{T}}^{(\text{AE})} = \left\{ \boldsymbol{x}^{(i)} | i = 1, \ldots, N_{\text{T}} \right\}, \tag{43}$$

$$\boldsymbol{x}^{(i)} = \left[x_1^{(i)} \cdots x_{n_1}^{(i)} \right]^{\mathsf{T}}. \tag{44}$$

Consider a three-layered autoencoder, which is the most basic structure of an autoencoder, as an example. The condition $n_3 = n_1$ is satisfied, and its loss function $L^{(\text{AE})}$ is given as follows:

$$L^{(\text{AE})} = \frac{1}{n_1} \left\| \boldsymbol{x}_3^{(i)} - \boldsymbol{x}^{(i)} \right\|^2 = \frac{1}{n_1} \sum_{m=1}^{n_1} \left(x_{3,m}^{(i)} - x_m^{(i)} \right)^2. \tag{45}$$

It means that the network is trained to make its output imitate the input. In the case of $n_2 < n_1$, the output of the hidden layer $\boldsymbol{x}_2^{(i)}$ can be used as the encoded data of the input $\boldsymbol{x}^{(i)}$, because even though the dimension of $\boldsymbol{x}_2^{(i)}$ is less than that of the input, $\boldsymbol{x}_3^{(i)}$ can be generated using $\boldsymbol{x}_2^{(i)}$.

Denoising Autoencoder

The DAE, which is also a kind of Reg-NN, has a structure similar to that of an autoencoder. The basic concept of the DAE is to train the network in order to estimate the original signal from the noisy input. The training dataset of a DAE can be expressed as follows:

$$D_{\mathrm{T}}^{(\mathrm{DAE})} = \left\{ (\boldsymbol{x}^{(i)}, \boldsymbol{s}^{(i)}) | i = 1, \ldots, N_{\mathrm{T}} \right\}, \tag{46}$$

$$\boldsymbol{x}^{(i)} = \boldsymbol{s}^{(i)} + \boldsymbol{n}^{(i)}, \tag{47}$$

where $\boldsymbol{n}^{(i)}$ denotes the i-th artificially generated noise sequence; therefore, $\boldsymbol{x}^{(i)}$ denotes the superposition of the original signal $\boldsymbol{s}^{(i)}$ and noise. In addition, the autoencoder $n_3 = n_1$ is satisfied. Let $\boldsymbol{y}^{(i)}$ denote the output of the DAE when $\boldsymbol{x}^{(i)}$ is the input. The loss function, L, of the DAE is defined as follows:

$$L = \frac{1}{n_1} \left\| \boldsymbol{y}^{(i)} - \boldsymbol{s}^{(i)} \right\|^2 . \tag{48}$$

The network is trained to estimate the original signal when the superposition of the signal and noise is input. Figure 1 depicts the original signal, noisy signal, and noise-reduced signal.

Ensemble Denoising Autoencoder

Here, we propose an E-DAE as a new model of noise reduction by using the NN. Figure 2 depicts a conceptual diagram of the E-DAE. It consists of independently trained M DAEs, namely, $\mathrm{DAE}_{\langle 1 \rangle}$, ..., and $\mathrm{DAE}_{\langle M \rangle}$. Let $\boldsymbol{y}_{\langle m \rangle}$ denote the output of $\mathrm{DAE}_{\langle m \rangle}$ when noisy data $\boldsymbol{x}$ is input. Subsequently, the output of E-DAE $\boldsymbol{y}$ is defined as follows:

$$\boldsymbol{y} = \frac{1}{M} \sum_{m=1}^{M} \boldsymbol{y}_{\langle m \rangle} . \tag{49}$$

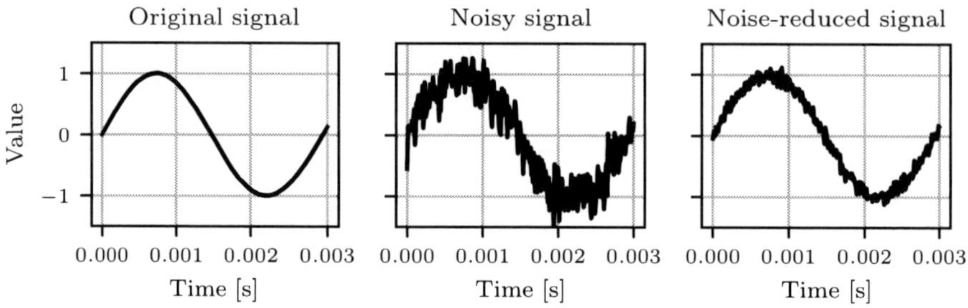

Figure 1. Example plots of original signal $\boldsymbol{s}$, noisy signal $\boldsymbol{x}$, and noise-reduced signal $\boldsymbol{y}$.

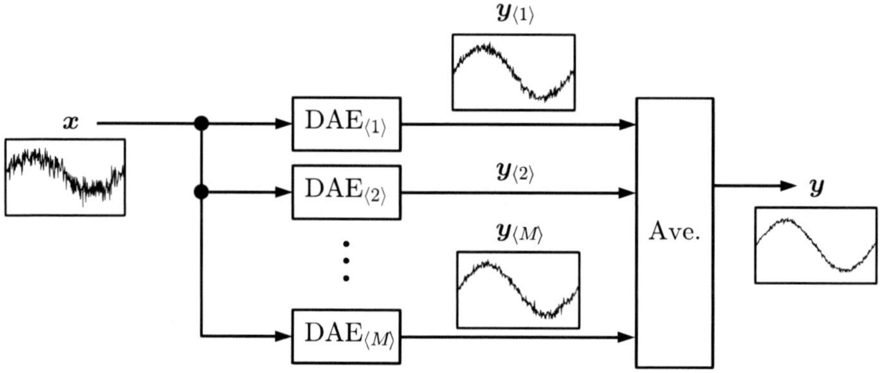

Figure 2. Concept of E-DAE.

Conversely, the output of the E-DAE is the mean of the outputs of DAEs. This ensemblification may enhance the robustness against noise.

Although, of course, we can consider other types of an ensemble in this step, such as a weighted mean or the stacking, we only focus on the simplest ensemble using the mean of the outputs. It is because the aim of this chapter is to introduce this new model and discuss its effectivity. It is important to start from the simplest model to discuss the effectivity and extensibility.

In this study, the following two strategies are examined to independently train each DAE:

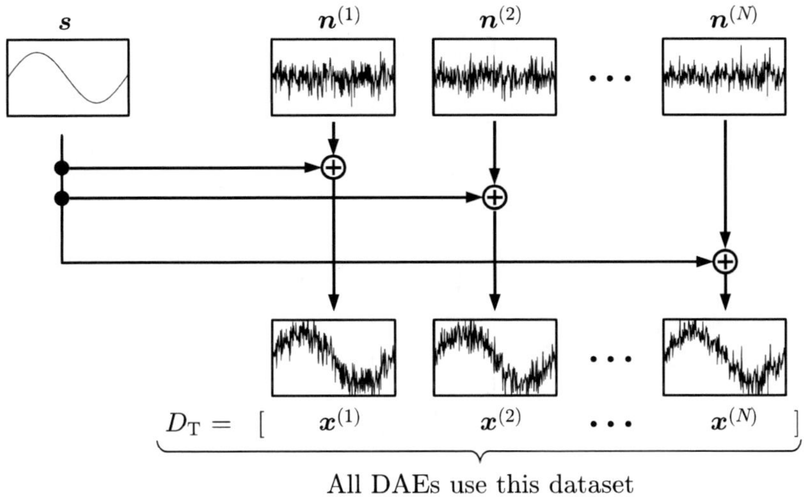

Figure 3. Preparation of the training dataset for E-DAE1.

1. The initial values of model parameters, namely, $\boldsymbol{W}$, $\boldsymbol{V}$, $\boldsymbol{b}$, and $\boldsymbol{c}$, of each DAE are different from one other
2. In addition to the initial values of model parameters, noise sequences used in the training dataset are different from one another

We hypothesized that it could enhance the robustness by using different noise sequences in the training of the different DAE because noise sequences appearing in the real data are different every time. The second strategy is introduced to examine the hypothesis of ours.

Hereinafter, we call the E-DAE based on the first strategy E-DAE1 and that based on the second one E-DAE2. Figures 3 and 4 depict how to prepare the datasets for E-DAE1 and E-DAE2, respectively, in a simple case that there is only one type of original signal. In the E-DAE1, all DAEs use the same training dataset D_{T}; hence we need to generate N realizations of noise sequence. In the E-DAE2, each DAE uses a different dataset. $D_{\mathrm{T}}^{(1)}$ in the figure 4 is a dataset used by $\mathrm{DAE}_{\langle 1 \rangle}$. In this case, we need to generate $N \times M$ realizations of noise sequence to prepare the datasets.

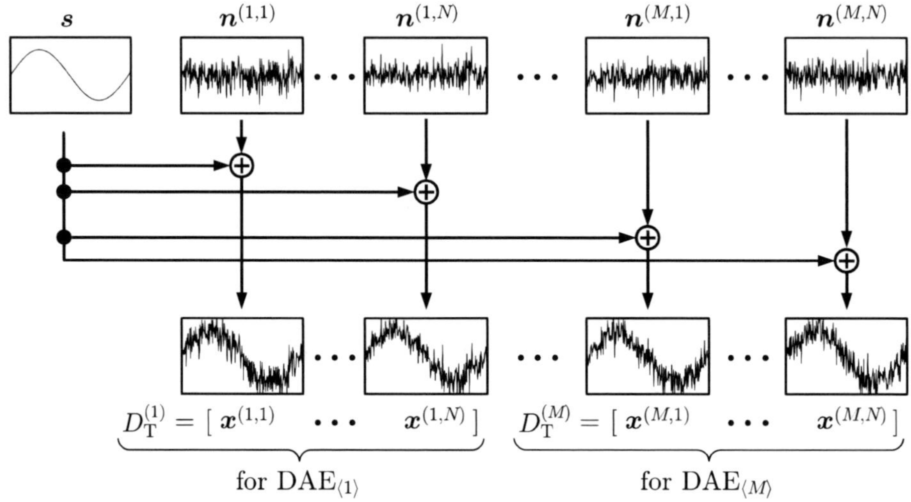

Figure 4. Preparation of the training dataset for E-DAE2.

SIMULATION OF E-DAE

We conducted some simulations to evaluate the performance of the E-DAE. In this section, the outline and results of the simulation are reported.

Dataset

We use the two signals, namely, $\boldsymbol{s}_1$ and $\boldsymbol{s}_2$, as the original signals for the simulation. We have the following:

$$s_1[k] = A \exp\left(-\left(\frac{(k-N/2)\Delta t}{\tau}\right)^2\right) \sin(2\pi f_c(k-N/2)\Delta t) + 0.5, \quad (50)$$

$$s_2[k] = 0.5. \quad (51)$$

Notably, $\boldsymbol{s}_1$ denotes a sine-Gaussian signal and $\boldsymbol{s}_2$ a bias signal. A sine-Gaussian signal is generally used as a model signal of a short-duration transient. Preparing these two signals for the simulation enables us to examine whether the E-DAE can extract this kind of short-duration transient and also, whether it can return a constant value if no signal is contained in the data. In this study, we set the parameters as follows:

$$A = 0.468, \quad \tau = 0.016\,\text{s}, \quad f_c = 300\,\text{Hz}. \tag{52}$$

The sampling interval, Δt, and data points, N, are set to $1/4096$ s and 300, respectively. The waveforms are depicted in Figures 5 and 6.

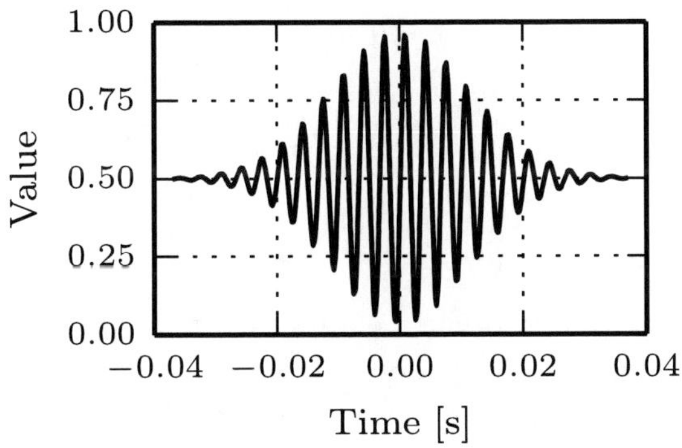

Figure 5. Original waveform s_1.

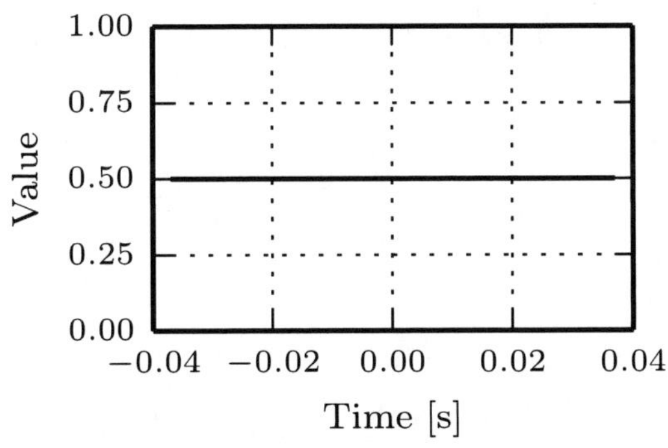

Figure 6. Original waveform s_2.

We use random numbers that follow Gaussian distribution with a mean of 0 and standard deviation of 0.15 for the noise sequences used for preparing the datasets.

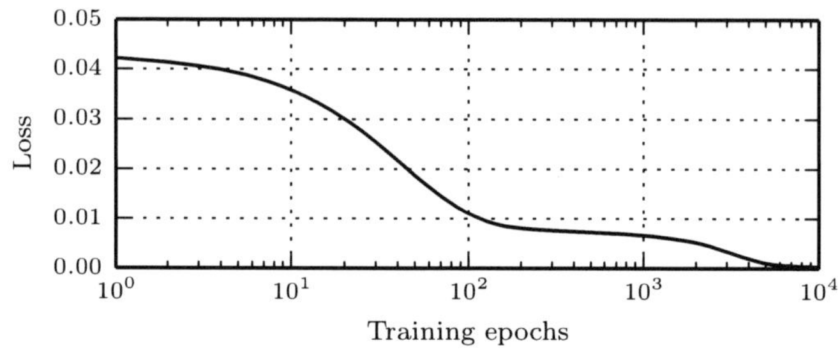

Figure 7. Learning curve of a DAE.

Model Configuration

The simulation aims to investigate the relevance of the number of ensembles to the performance of both E-DAE1 and E-DAE2 and to reveal whether there is a significant difference between the performances of E-DAE1 and E-DAE2. To construct E-DAE1 and E-DAE2, we prepare 20 DAEs with the same number of units, $n_1 = n_3 = 300$ and $n_2 = 200$. The initial values of the model parameters, namely, $\boldsymbol{W}$, $\boldsymbol{V}$, $\boldsymbol{b}$, and $\boldsymbol{c}$, of each DAE are given by random numbers, which follow Gaussian distribution with a mean of 0 and standard deviation of 0.1. For both the cases of E-DAE1 and E-DAE2, a DAE with the same ID uses the same initial values. Conversely, the model parameters of DAE_1 in E-DAE1 and those of DAE_1 in E-DAE2 are initialized using the same random numbers. Subsequently, we employ the mini-batch gradient descent method with a batch size of 32 for the learning process, and the learning iterates 10,000 times.

Results

The learning curve, which shows how the performance of the NN improves with the number of learning iterations, of one of the DAEs, is depicted in Figure 7. We can confirm that the loss is sufficiently converged and that the noise contained in the data is reduced using the DAE.

Figure 8 depicts the results by using E-DAE1 with the number of DAEs as 1, 2, 4, and 20. It is evident the noise reduction achieved by E-DAE1 improves upon using more DAEs. Let $\boldsymbol{y}^{(i)}_{\langle m \rangle}$ denote the result of noise reduction on the validation data, $\boldsymbol{x}^{(i)}$, by using $\text{DAE}_{\langle m \rangle}$. We define the mean loss, L_M, as

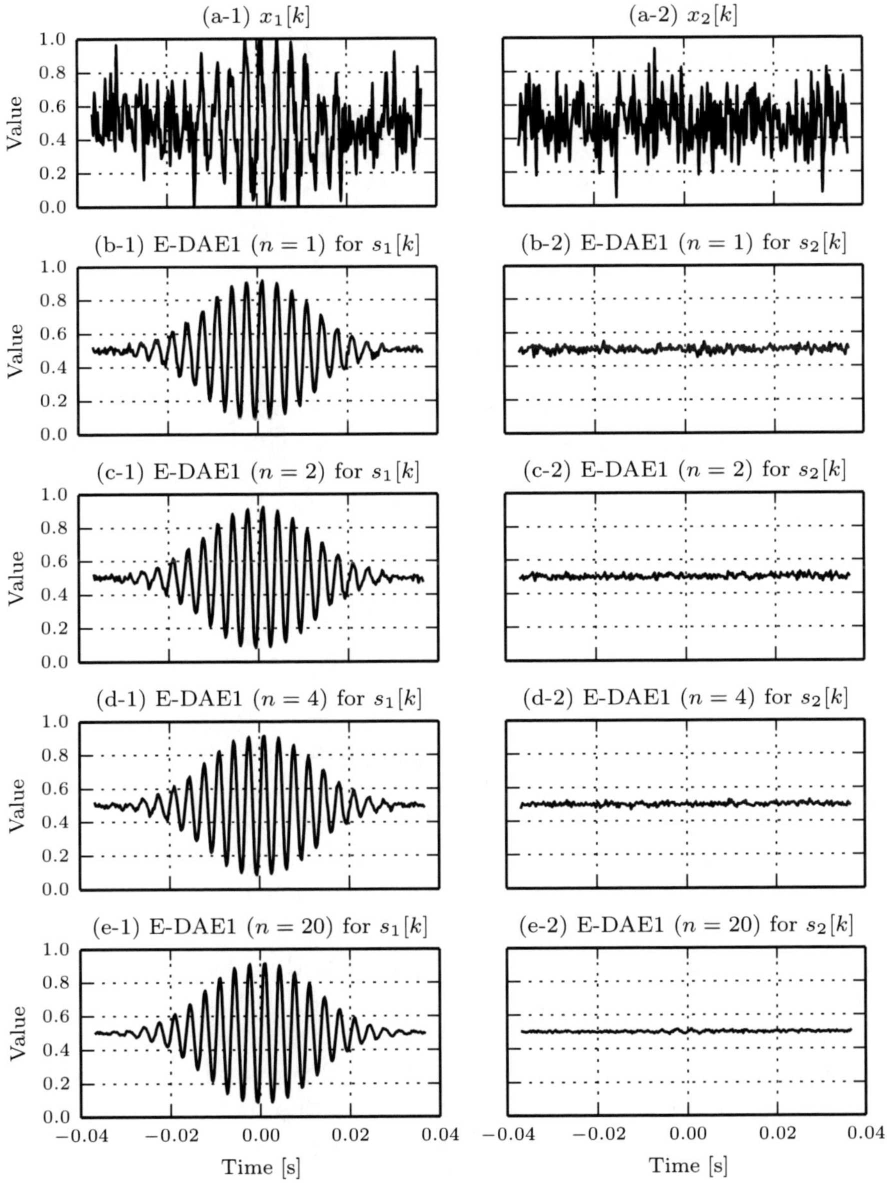

Figure 8. Results using E-DAE1.

follows:

$$L_M = \frac{1}{N_{\mathrm{T}}} \sum_{i=1}^{N_{\mathrm{T}}} L_M^{(i)}, \tag{53}$$

$$L_M^{(i)} = \frac{1}{{}_{20}\mathrm{C}_M} \sum_{j=1}^{{}_{20}\mathrm{C}_M} L_{M,\boldsymbol{m}_{M,j}}^{(i)}, \tag{54}$$

$$L_{M,\boldsymbol{m}}^{(i)} = \frac{1}{M} \sum_{k=1}^{M} L_{\langle m_k \rangle}^{(i)}, \tag{55}$$

$$L_{\langle m \rangle}^{(i)} = \frac{1}{n_1} \left\| \boldsymbol{y}_{\langle m \rangle}^{(i)} - \boldsymbol{s}^{(i)} \right\|^2, \tag{56}$$

where $\boldsymbol{m}_{M,j}$ consists of the j-th possible combination of M integers among the 20 integers;

$$\begin{aligned} \boldsymbol{m}_{M,j} = \{m_1, m_2, \ldots, m_M\} \subset \{1, 2, \ldots, 20\}, \\ \boldsymbol{m}_{M,j} \neq \boldsymbol{m}_{M,k} \quad (j \neq k). \end{aligned} \tag{57}$$

$L_{M,\boldsymbol{m}_{M,j}}^{(i)}$ is the mean of the loss $L_{\langle m \rangle}^{(i)}$ over a specific combination of M DAEs from the prepared 20 DAEs, and $L_M^{(i)}$ denotes the mean of $L_{M,\boldsymbol{m}_{M,j}}^{(i)}$ over the all possible combinations. Figure 9 depicts the relationship between the mean loss, L, and the number of DAEs used in E-DAE1. It is evident that the mean loss decreases upon increasing the number of DAEs used in E-DAE1. Therefore, the effectiveness of ensemble learning in noise reduction is confirmed.

Along with the simulation of E-DAE1, we conducted a simulation of E-DAE2 as well, following which we compared the results for both. The top two rows in Figure 10 depict the outputs of E-DAE1 and E-DAE2 upon inputting the same validation data. Both the cases of $\boldsymbol{s}_1$ and $\boldsymbol{s}_2$ are depicted. The outputs of E-DAE1 and E-DAE2 appear to be identical when looked at with naked eyes. The differences between the outputs are depicted in the bottom row in Figure 10; there is only a difference with an order of 10^{-5}. Figure 11 depicts the mean loss of both E-DAE1 and E-DAE2; it also indicates that there is no significant difference between the performances of both.

The performance of the E-DAE might depend on how independently each DAE is trained. According to the simulation results, it is sufficient to initialize the model parameters independently, whereas using different noise sequences

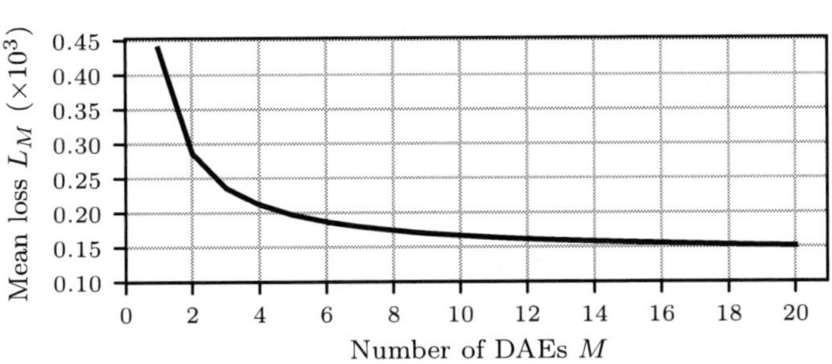

Figure 9. Mean loss and number of ensembles.

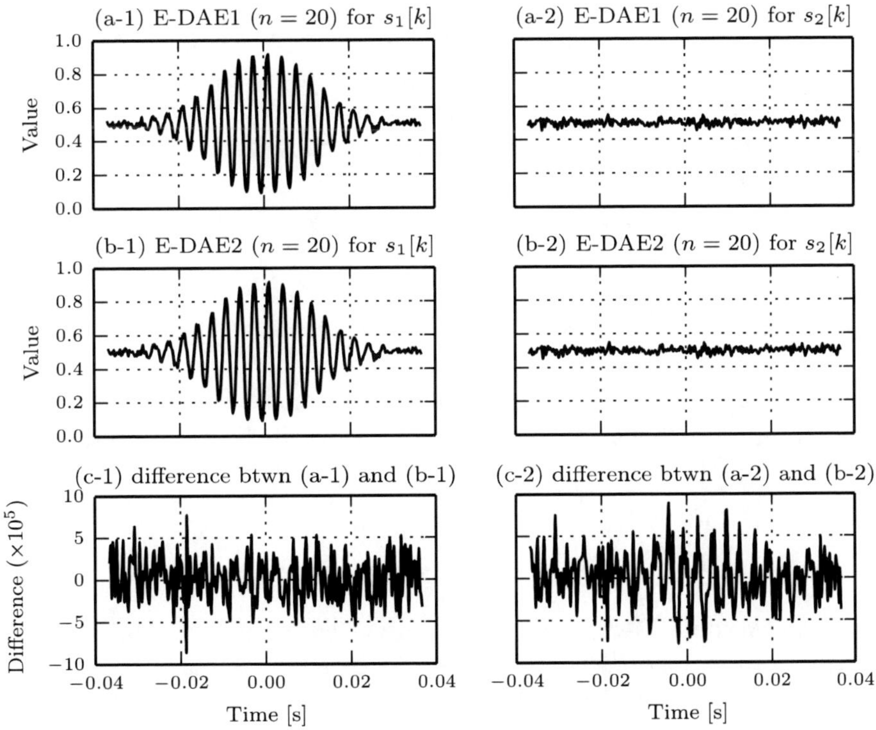

Figure 10. Results using E-DAE1 and E-DAE2, and the difference between their outputs.

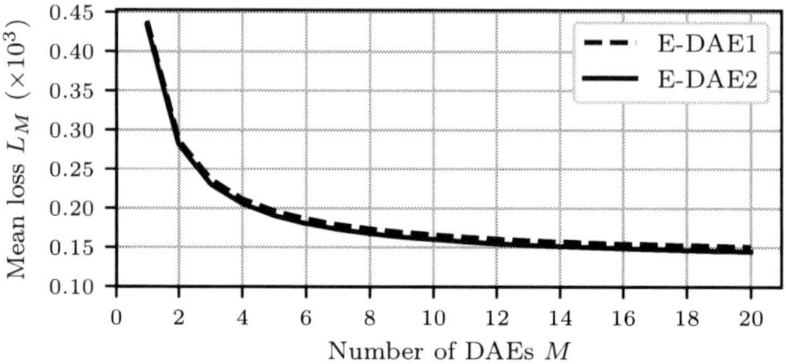

Figure 11. Mean loss and number of ensembles.

for making the training dataset has an insignificant influence. It can be explained as follows. On the basis of the formula of parameter learning in Eq. (14), each parameter is updated to the nearest minimum of the loss function along the gradient. Because the loss is defined with the correct values, the gradient does not change considerably when the noise sequence used in the training dataset changes. However, when the initial values of the model parameters change, the nearest minimum also changes. Therefore, the independent initialization of the model parameters generates independence between models.

On the basis of the discussion, it is confirmed that the E-DAE achieves better performance by making DAEs more independent. In addition, it is assumed that using different probability distributions to generate initial values for each DAE, or using a different network architecture for each DAE, can be the possible extensions of the E-DAE.

CONCLUSION

We introduced the E-DAE as an application of ensemble learning in order to perform a noise-reduction task. We briefly reviewed the NN and its related topics as the fundamentals of the E-DAE, introduced the definition of the E-DAE, and reported its simulation-based evaluation. As a result of the simulation, it was confirmed that the ensemblification of DAE improves its noise-reduction performance, and that the performance improved as the number of DAEs contained in the E-DAE increased. Furthermore, we revealed the fact that independent initialization of the model parameters is sufficient for achieving better performance.

This study can be extended to various directions. For example, applications in DNNs, and using statistical processing other than averaging. Furthermore, we plan to improve the method and utilize it to gravitational-wave-data analysis.

ACKNOWLEDGMENTS

This work was supported in part by JSPS Grant-in-Aid for Young Scientists (B) (Grant No. 17K13179; Y. Omae), JSPS Grant-in-Aid for Young Scientists (Grant No. 19K14717; K. Sakai and Grant No. 19K20062; Y. Omae), JSPS Grant-in-Aid for Scientific Research (C) (Grant No. 17K05437 and Grant No. 19K12227; H. Takahashi and Grant No. 20K12090; Y. Omae and H. Takahashi), JSPS Grant-in-Aid for Scientific Research (B) (Grant No. 19H01901; H. Takahashi) and JSPS KAKENHI (Grant No. 17H06358; H. Takahashi).

REFERENCES

[1] Vincent, P., Larochelle, H., Bengio, Y. & Manzagol, P. A. (2008). Extracting and Composing Robust Features with Denoising Autoencoders. *Proceedings of the 25th international conference on Machine learning*. 1096-1103.

[2] Rumelhart, D. E., Hinton, G. E. & Williams, R. J. (1986). Learning Representations by Back-propagating Errors. *Nature*. 323, 533-536.

[3] Adeli, H. (2001). Neural networks in Civil Engineering: 1989-2000. *Computer-Aided Civil and Infrastructure Engineering*. 16(2), 126-142.

[4] Christodoulou, C. & Georgiopoulos, M. (2000). *Applications of Neural Networks in Electromagnetics*. Artech House, Inc..

[5] Yagawa, G. & Okuda, H. (1996). Neural Networks in Computational Mechanics. *Archives of Computational Methods in Engineering*. 3(435).

[6] Hu, Y. H. & Hwang, J. N. (Eds.). (2002). *Handbook of Neural Network Signal Processing*. CRC Press.

[7] Russakovsky, O., Su, H., Krause, J., Satheesh, S., Ma, S., Huang, Z., Karpathy, A., Khosla, A., Bernstein, M., Berg., A. C. & Fei-Fei, Li. (2014). ImageNet Large Scale Visual Recognition Challenge. *International Journal of Computer Vision*. 115(3).

[8] Lu, X., Tsao, Y., Matsuda, S. & Hori, C. (2013). Speech Enhancement based on Deep Denoising Autoencoder. *Proceedings of INTERSPEECH 2013*.

[9] Lu, X., Tsao, Y., Matsuda, S. & Hori, C. (2014). Ensemble Modeling of Denoising Autoencoder for Speech Spectrum Restoration. *Proceedings of INTERSPEECH 2014*.

[10] Xiong, P., Wang, H., Liu, M. & Liu, X. (2015). Denoising Autoencoder for Eletrocardiogram Signal Enhancement. *Journal of Medical Imaging and Health Informatics*. 5(8), 1804-1810.

[11] Zhang, H., Weng, T. W., Chen, P. Y., Hsieh, C. J. & Daniel, L. (2018). Efficient Neural Network Robustness Certification with General Activation Functions. *Advances in Neural Information Processing Systems*. 4939-4948.

[12] Cheung, R. K. M, Lustig, I. & Kornhauser, A. L. (1990). Relative Effectiveness of Training Set Patterns for Backpropagation. *Proceedings of IEEE International Joint Conference on Neural Networks*. 6, 673-678.

[13] Szu, H. & Hartley, R. (1987). Fast Simulated Annealing. *Physics Letters A*. 122, 157-162.

[14] Qian, N. (1999). On the Momentum Term in Gradient Descent Learning Algorithms. *Neural Networks*. 12(1), 145-151.

[15] Hastie, T., Tibshirani, R. & Friedman, J. H. (2009). *The Elements of Statistical Learning: Data Mining, Inference, and Prediction*. Springer Science & Business Media, NY, USA.

[16] Panchal, G., Ganatra, A., Shah, P. & Panchal, D. (2011). Determination of Over-learning and Over-fitting Problem in Back Propagation Neural Network. *International Journal on Soft Computing*. 2(2), 40-51.

[17] Mehta, P., Bukov, M., Wang, C. H., Day, A. G., Richardson, C., Fisher, C. K. & Schwab, D. J. (2019). A High-bias, Low-variance Introduction to Machine learning for Physicists. *Physics Reports*. 810, 1-124.

[18] Sutskever, I., Martens, J., Dahl, G. & Hinton, G. (2013). On the Importance of Initialization and Momentum in Deep Learning. *Proceedings of the 30th International Conference on Machine Learning*. 28(3), 1139-1147.

[19] Zhang, S., Choromanska, A. E. & LeCun, Y. (2015). Deep Learning with Elastic Averaging SGD. *Proceedings of the 28th International Conference on Neural Information Processing Systems*. 1, 685-693.

[20] Omae, Y., Kon, Y., Kobayashi, M., Sakai, K., Shionoya, A., Takahashi, H., Akiduki, T., Nakai, K., Ezaki, N., Sakurai, Y. & Miyaji, C. (2017). Swimming Style Classification Based on Ensemble Learning and Adaptive Feature Value by Using Inertial Measurement Unit. *Journal of Advanced Computational Intelligence and Intelligent Informatics*. 21(4), 616-631.

[21] Drucker, H. (1997). Improving Regressors Using Boosting Techniques. *Proceedings of the Fourteenth International Conference on Machine Learning*. 107-115.

About the Authors

Kazuki Sakai is currently an Assistant Professor of the Department of Electronic Control Engineering, National Institute of Technology (KOSEN), Nagaoka College, Japan, from April 2018. He received his Doctor degree in Engineering from the Department of Information Science and Control Engineering, Nagaoka University of Technology, Japan, in 2018. His main research interest is the application of the Adaptive Data Analysis Methods into the Time-Series Signal Processing, especially Gravitational Wave Data Analysis. He is a member of The Japanese Society for Artificial Intelligence (JSAI), the Institute of Electronics, Information and Communication Engineers (IEICE), the Physical Society of Japan (JPS), and so on.

Yuto Omae is currently an Assistant Professor of Department of Industrial Engineering and Management, College of Industrial Technology, Nihon University, Japan from April 2019. He received his Doctor degree in Engineering from the Department of Information Science and Control Engineering, Nagaoka University of Technology, Japan in 2016. His research interests are the Adaptive Data Analysis Methods, which include Machine Learning and Artificial Intelligence, to various data (sensor data and educational data etc.). He is a member of Institute of Electrical and Electronics Engineers (IEEE), The Japanese Society for Artificial Intelligence (JSAI), the Institute of Electronics, Information and Communication Engineers (IEICE) and Japan Society for Fuzzy Theory and Intelligent Informatics (J-SOFT) etc.

Hirotaka Takahashi is currently a Professor of the Graduate School of Integrative Science and Engineering, Tokyo City University, Japan from September 2020. He received his Doctor degree in Science from the Graduate School of Science and Technology, Niigata University, Japan in 2005. He was a PD at Max Planck Institute for Gravitational Physics (Albert Einstein Institute) in Germany from September 2005 to August 2007. He was a faculty staff member at various universities. His research interests include the Adaptive Data Analysis Methods, Information Theory and Gravitational Wave Physics and Astronomy (General Relativity) etc. He is a member of the Institute of Electronics, Information and Communication Engineers (IEICE) and the Physical Society of Japan (JPS) etc.

In: An Introduction to Approaches … ISBN: 978-1-53618-680-2
Editor: Yi-Tung Chan

Chapter 3

COMPUTER-AIDED DIAGNOSIS SYSTEM FOR BONE FRACTURE DETECTION AND CLASSIFICATION: A REVIEW ON DEEP LEARNING TECHNIQUES

Leonardo Tanzi[1,*], Enrico Vezzetti[1], Alessandro Aprato[2], Andrea Audisio[2] and Alessandro Massè[2]

[1]Department of Management and Production Engineering, Politecnico di Torino, Torino, Italy

[2]School of Medicine, University of Turin, Torino, Italy

ABSTRACT

Bone fracture detection and classification was a large discussed topic over the last few years and many researchers proposed different technological solutions to tackle this task. Despite this, a universal approach able to support the classification of fractures in the human body

* Corresponding Author's Email: leonardo.tanzi@polito.it.

still does not exist today. We aim to provide a first discussion through a selection of research works done in the technological domain, with a specific focus on Deep Learning. The objective is to stimulate a knowledge improvement in bone fracture classification, necessary to start the development of an optimal shared framework. The evaluation phase involved a first qualitative assessment based on strengths and limitations, providing a usage scenario evaluation. This could support the development of a Computer-Aided Diagnosis (CAD) system able to help doctors in reducing diagnosis time while increasing their reliability and accuracy in complex cases and during stressful working conditions.

Keywords: deep learning, neural networks, bone fracture, CAD system

1. Introduction

In the last decades, several medical procedures have improved year by year with the help of technology [1, 2]; among these, great progress has been made in the orthopedic field. Bones' fractures are one of the most common injuries nowadays. Every year 2.7 million fractures occur across the EU6 nations, France, Germany, Italy, Spain, Sweden, and the UK [3], an incredible number of people suffer for this disorder and the implications of an untreated fracture may lead to permanent damages or even death. A great responsibility in this lies with the doctors, who have to evaluate tens of X-Ray images a day. The technology utilized for first diagnosis is mostly X-Ray, which is a modality used for more than one hundred years and is still frequently exploited. It is challenging for doctors to evaluate X-Ray images: firstly, X-Ray could hide certain particularities of bone; secondly, a long experience is needed to correctly classify different types of fractures; thirdly, doctors often have to act in emergency situations and may be constrained by fatigue. Actually, it has been shown that the performance of radiologists in the interpretation of musculoskeletal radiographs decreases in fracture detection at the end of the work day compared to the beginning of the shift [4]. In addition, radiographic interpretation often takes place in environments without the availability of qualified colleagues for second opinions [5]. The success of the treatment and prognosis strongly depends on an accurate classification of the fracture among standard types, such as those defined by the Arbeitsgemeinschaft fu¨r Osteosynthesefragen (AO-foundation). But studies show a relative under-use of this comprehensive classification system [6], perhaps due to fast-paced working conditions in particular in emergency departments. In clinical practice, this phenomenon compromises the evidence-based management of patients, while the lack of uniform records affects future retrospective studies. Computer-aided diagnostic systems support physicians in the interpretation of medical images. The last papers published on this topic suggest how the implementation of Deep Learning protocols can outperform physicians' diagnostic outcomes. For this reason we decided to focus on this paradigm, underlying the main aspects behind Deep Learning application in the next section.

2. Deep Learning Scenario

Unfortunatly at present a univoque and shared methodology in bone fracture classification is still not available, but many different studies are proposed especially in the domain of machine learning. Machine Learning is generally defined as the practice of using algorithms to parse data, learn from it, and then make a prediction about something in the world. Two types of machine learning can be defined (we do not consider reinforcement learning for the sake of simplicity): supervised, where the network learns from labelled data, and unsupervised, where the computer learns by itself without any help from the labelling. Deep Learning is becoming more and more widely used in the world of computer vision technologies, giving astonishing results in different fields of application, for example surgery [7] and face recognition [8]. A neural network, i.e., the classic architecture used in Deep Learning application, is composed of input, hidden and output layers, all of which are composed of nodes. All these subsequent layers define a function with thousands or even millions of parameters (called weights and biases). The input layer takes in a numerical representation of data (e.g., images with pixel specs), the output layer output predictions, while the hidden layers are correlated with most of the computation. In a typical classification problem, the network is fed with the input images and tries to assign a specific pre-defined class to each of them. After the prediction, it calculates a loss function (how much the predicted classes differ from the original) and adjust its parameters with gradient descent and backpropagation. The images are fed different times through the network back and forth, until the network gets a certain accuracy. In order to be clear, we are going to define some basic Deep Learning glossary.

- *Train, Validation, Test Set*. It is a common procedure to divide the dataset in 3 different groups: Train Set, to train the network, Validation Set, to test the network after each epoch and Test Set, to test the network at the end of the training with images that has never seen before.
- *Transfer Learning* It has been demonstrated that the parameters of a network trained on a specific task using a dataset X, may be re-used to adapt the network to solve a different task using a dataset Y, instead of using randomly initialized parameters.
- *Data augmentation* Data augmentation is a strategy that enables to significantly increase the diversity of data available for training models, without actually collecting new data, using techniques such as cropping, padding, and flipping. Some examples are showed in Figure 1.
- *Convolutional Neural Networks (CNN)* CNN are the most used technology in computer vision as they are able to successfully capture the spatial and temporal dependencies in an image through the application of relevant filters, saving memory space and limiting the number of parameters to compute. The input images are made to convolute with different filters to extract features and then passed to pooling layer to lower the size of data. After this process the output is passed to the fully connected layers as showed in Figure 2.

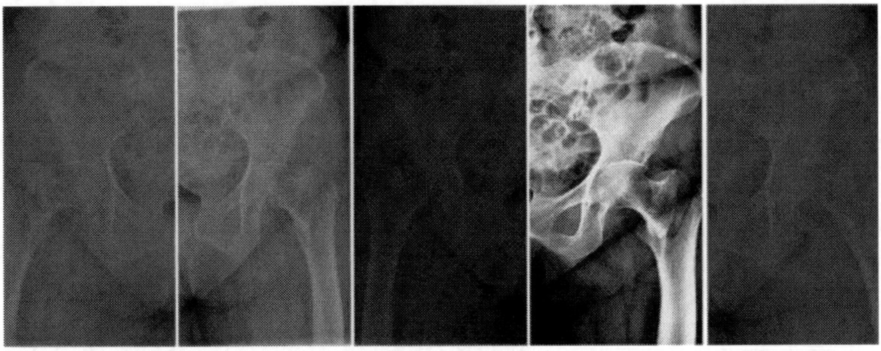

Figure 1. Examples of data augmentation techniques.

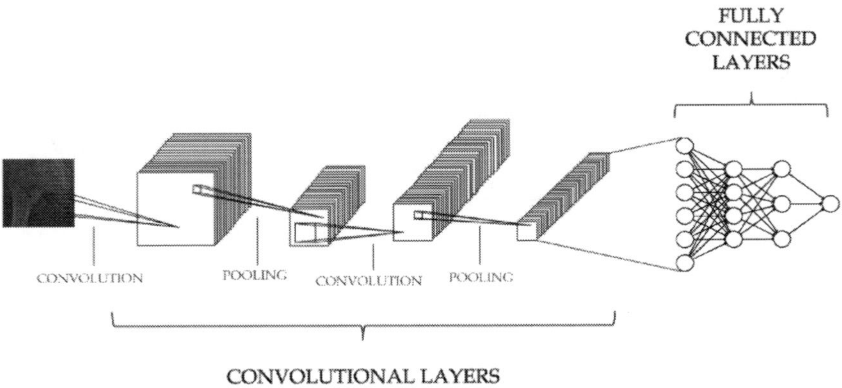

Figure 2. Convolutional Neural Network architecture. Convolutional layers are the layers where filters are applied to the original image, or to other feature maps in a deep CNN. Pooling layers' function is to progressively reduce the spatial size of the representation to reduce the number of parameters and computation in the network.

3. METHODOLOGY

The review process proposed in the present paper was based on a desktop research analysis [9] using Google Scholar, MEDLINE and Embase, run on a selection of specific papers of the specific research domain moving from basic approaches to the main advanced solutions. Initial prior works for detection and classification of fractures [10–12] have focused on conventional machine learning processes consisting of pre-processing, feature extraction and classification steps. Recently, impressive results have been obtained using Deep Learning [13] methods. We summarized a brief overview about Deep Learning in section 2. A total of 237 records were identified through database searching. 107 records were screened and 63 of them excluded, resulting in 44 full-text articles assessed for eligibility. Among these papers, we selected 10 records for analysis. We excluded 21 studies using surpassed technologies and 13 studies tackling non-ineherent arguments. The majority of them pursue the classification between fractured and not fractured bones, while just two of them tried to classify the different types of fractures. We have chosen papers which, in our personal opinion, contain strengths given by a Deep Learning approach that should be used in order to develop a generic tool able to classify every type of fracture in each bone of the human body.

4. APPROACHES

To the best of our knowledge, a work that tries to define an ideal method to classify fractures valid for each bone in the human body does not exist. In this paper, we aim to evaluate different papers and show their assets that can contribute in the definition of a future baseline approach. Before the advent of CNN, the pre-processing phase was a fundamental part of the work. For example, in the work of Dimililer [14], the author's aim was to classify whether a bone in an X-Ray image is fractured or not. The system is composed by a neural network following a pre-processing phase. The tool has been trained with 30 images and tested with 70. The images contain different fractures in size and illumination conditions for each subject. In the pre-processing phase, the images are processed using techniques such as Haar Wavelet and Scale-Invariant Feature Transform (SIFT). Haar Wavelet transform is needed to pre-process images in order to compress them and save memory space, SIFT is a powerful method to detect feature points with high resilience to several issues like rotation, compression, and scaling. In the classification phase the author implements a 3-layers neural network with 1024 input neurons. The whole pipeline is showed in Figure 3.

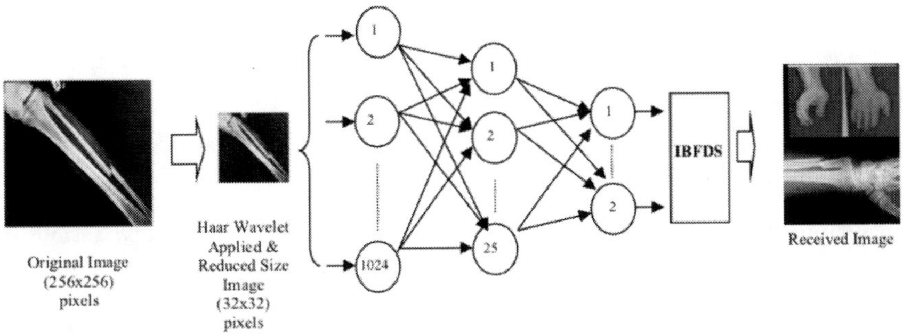

Figure 3. Pipeline of the architecture proposed by Dimililer [14].

This paper describes the technology used before the advent of the convolutional layers. The pre-processing phase was a fundamental part of the process in order to feed the fully connected layers with the correct information. All the remaining papers use convolutional layers to extract features before feeding the fully connected layers.

A dataset of 100 images is quite scant for Deep Learning applications: to solve this problem, Kim and MacKinnon [15] proposed the implementation of transfer learning techniques in order to classify wrist fractures in two classes: broken and unbroken. According to the authors, this was the first work were transfer learning from pre-trained CNNs has been successfully applied to the problem of fracture detection on plain radiographs. The final dataset was composed of 1.389 images, 695 wrist radiographs showing a fracture and 694 showing no fracture. Subsequently, data augmentation was applied: horizontal flip, rotation (between 0° and 25°), width and height shift (by a factor of 0-15%), shearing (between 0-10%) and zoom (between 0-15%). This resulted in an overall amplification by a factor of 8 and 5.560 images in the fracture group and 5.552 images in the no fracture group. This classification was checked and verified by a radiology registrar with 3 years' radiology experience. The network used from this purpose was InceptionV3 [16] originally trained with ImageNet [17] dataset and then adapted and re-trained for the broken/unbroken classification, modifying the top-layer of the network. In this case, the authors demonstrate how transfer learning could be applied with a dataset of images completely not relatable, i.e., ImageNet.

At contrary, Lindsey et al. [18] pre-trained their network with a large dataset of bones images. The aim of their work was to implement a tool that could help doctors in diagnosis, in order to distinguish if a wrist bone was fractured or not and which part of the bone was involved. The dataset consisted of 135.845 radiographs of a variety of body parts. Of these, 34.990 radiographs (Training Set) were posterior–anterior or lateral wrist views. The remaining 100.855 radiographs (Pre-Training Set) belonged to 11 other body parts: foot, elbow, shoulder, knee, spine, femur, ankle, humerus, pelvis, hip, and tibia. Every train image was labelled with a bounding box drawn by a group of senior orthopedic surgeons specialized in trauma. The model was a deep CNN, whose architecture is an extension of the common U-Net [19] model. The CNN has two outputs: the probability that the radiograph has a visible fracture and a heat map indicating for each location in the image the probability that the fracture is present in that location. The training of the model can be divided into two stages. In the first stage the model was pre-trained on the Pre-Training Set. In the second stage, the obtained model was fine-tuned using the Training Set, to specialize it to the task of detecting and localizing wrist fractures. After the training and testing phases of the CNN, the authors ran a controlled experiment with 40 emergency medicine clinicians, to evaluate each clinician's ability to detect fractures in wrist radiographs both with and without the help of the system. With the use of the proposed system, the clinicians average sensitivities and specificity improved. This study showed that specialists evaluation may be improved with the use of this system. This procedure should be applied in each work aiming at showing that a CAD system could help humans in evaluation. Pre-training the model before training it with the wrist bones images seem like a good procedure to adjust the parameters for the task, instead of using weights taken from a network that has been trained with a completely different dataset, e.g., ImageNet.

Unfortunately, not always you can count on large dataset to pre-train your network: Rajpurkar et al. [20] solved this problem with the introduction of MURA, one of the largest public radiographic images datasets. MURA dataset contains 14.863 musculoskeletal studies of the upper extremity. Each study contains one or more images taken from different views, with a total of 40.561 images, and was manually labelled by radiologists as normal or abnormal. The studies are divided in 9.045 normal and 5.818 abnormal for 7 different extremities including the shoulder, humerus, elbow, forearm, wrist, hand, and finger. The total number of multi-view images is 40.561. The dataset used by the authors is freely available, which implies that may be used by other researchers to pre-train a model for bones classification that can rely only on a small dataset.

Another good practice is to test different networks before choosing the one that gave the best performance, for example in the work of Olczak et al. [21]. The aim of this study was to assess if standard Deep Learning networks can be trained to identify if a bone is fractured or not in orthopedic radiographs. The dataset was composed of 256.458 hand, wrist, and ankle radiographs, with associated radiologist reports. The authors selected 5 common deep networks for this task: BVLC Reference CaffeNet network (8 layers), VGG CNN S network (8 layers), VGG CNN (16 and 19 layers' networks) and Network-in-network (14 layers). The networks were pre-trained on the ImageNet dataset and the last fully connected layer was replaced in order adapt the network for this specific task. As VGG16 had the best performance in the fracture class, the authors selected it for manual review. When comparing the network with the two senior orthopedic surgeons they found that the network performed similarly as the humans. Testing different existing networks and chose the one that performs best is a good practice in the field of neural network. The dataset contained a really high number of images, the highest among the datasets used in the papers we reviewed. This is obviously one of the most fundamental aspect when working with Deep Learning. With a huge dataset is not easy to label the images manually, and that is why the authors decided to label them automatically from the hospital information. This procedure is subject to errors and a second review may be useful.

Until now we discussed works which obtain good results working with large dataset: Yahalomi et al. [22] proposed a new technique to avoid this. The authors trained a Faster R-CNN [23], a machine vision neural network for object detection, to identify and locate distal radius fractures.

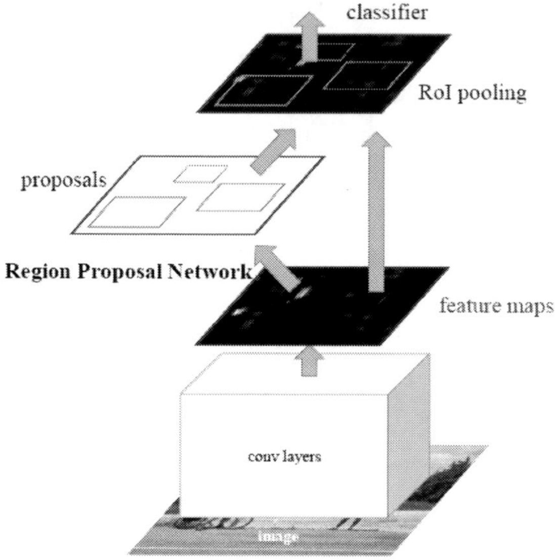

Figure 4. Faster-R-CNN architecture [23].

The initial dataset was composed by just 55 images of distal radius fractures and 40 images of hands without fractures. In addition, 25 images not showing hand bones are used for the negative test set. Each image was labelled with bounding boxes around the fractured area. The authors used a Faster R-CNN to achieve two different tasks: classifying weather the fracture is present or not and finding the fracture's location. Faster R-CNN is an evolution of R-CNN and Fast R-CNN where region proposals are generated by CNNs rather than using selective search. Faster R-CNN have 3 different phases. At first, the input images go through a CNN that extract feature maps. Secondly, a RPN (Region Proposal Network) is used for generating region proposal i.e., to pre-check which location contains an object without classifying the entity of the object. The output is then passed through a ROI (Region Of Interest) pooling to perform max pooling on inputs of non-uniform sizes and obtain fixed-size feature maps. Finally, the pooled area goes through CNN and two fully connected branches for class softmax and bounding box regressor, in order to detect the object class and returning the bounding box of that object. The neural network used in this work was VGG16. This is the only work we found that implements the technology of R-CNN to not only classify fracture but also detect the exact region of the fracture with a high accuracy (the results are demonstrated to be significantly more accurate than the detection achieved by physicians and radiologists).

Another improvement that can be applied is to select some specific region of the images to improve the network performance, as in Thurston et al. [24]. The aim of this paper was to increase the performance of the system described and already discussed from Kim & MacKinnon [15]. The highlight of this work is given by removing unnecessary parts of the image with semi-automated cropping process. The region of interest was defined using the Python OpenCV [25] *matchTemplate()* function. The method takes a template image and slides it across every position in the subject image (the wrist radiograph), returning the position in which the closest match was calculated. In this study, the region of interest was the distal radius. The template was, therefore, an anatomical representation of the distal radius. The template was produced by using a representative lateral wrist radiograph and applying a smoothing algorithm followed by a binary threshold to segment the bone. The process is showed in Figure 5.

A scaled template matching approach was adopted to account for different wrist sizes. The accuracy of the model was improved when the focus on the region of interest was applied. This extension study has demonstrated that the accuracy of the network to predict fractures can be increased by removing surplus imaging data. The process is still semi-automated, it should be possible to make it fully automated using different techniques, for example using feature matching that tries to match features between the template and the image.

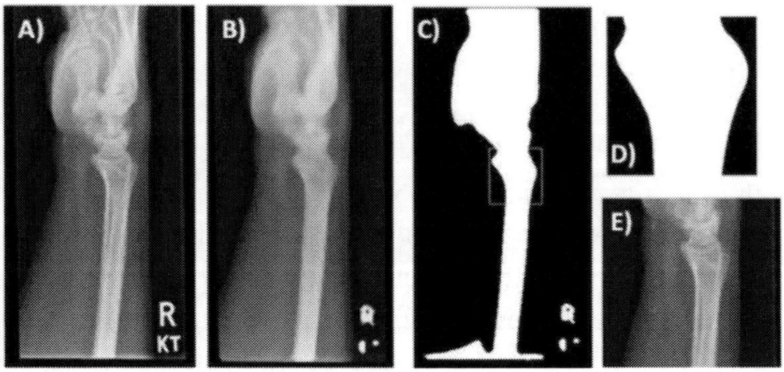

Figure 5. Pre-processing phases in [26].

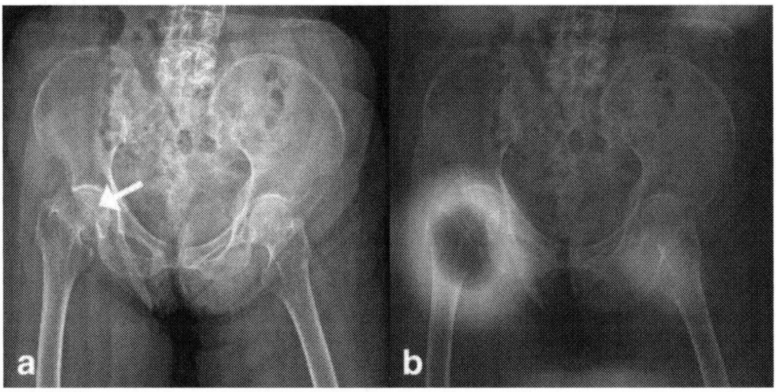

Figure 6. Heatmap generated by Grad-CAM [27].

Neural networks are often defined as "black boxes", as it is really hard to understand the learning process. This can be avoided with some visualization techniques, such as the one used by Cheng et al. [27]. The aim of this work is to use a CNN to classify and localize hip fractures on plain frontal pelvic radiographs. The localization phase is implemented by the use of gradient-weighted class activation mapping or Grad-CAM [28] to confirm the validity of the model. The authors used DenseNet [29] network for the classification task. To demonstrate that the CNN is actually focusing on the right area of the images, the authors implemented Grad-CAM to generate a heat map (Figure 6) in the images that the network classified as fractured. The heat maps computed with Grad-CAM were reviewed: after analyzing 49 heat map images, only two images identified the wrong activation site.

The use of Grad-CAM can confirm that the network is actually focusing on the correct area of images.

Since now, we just dealt with paper tackling with binary classification between broken and unbroken bones. The first work concerning the classification in different fractures was proposed by Chung et al. [30]. This work addressed the problem of classification in different types of fractures in the proximal humerus bone. To evaluate the performance of fracture classification, the authors refers to Neer's classification, which is the most commonly used classification for the proximal humerus fracture and distinguish 4 different types of fracture: greater tuberosity (B), surgical neck (C), 3-part (D), and 4-part (E) (Figure 7).

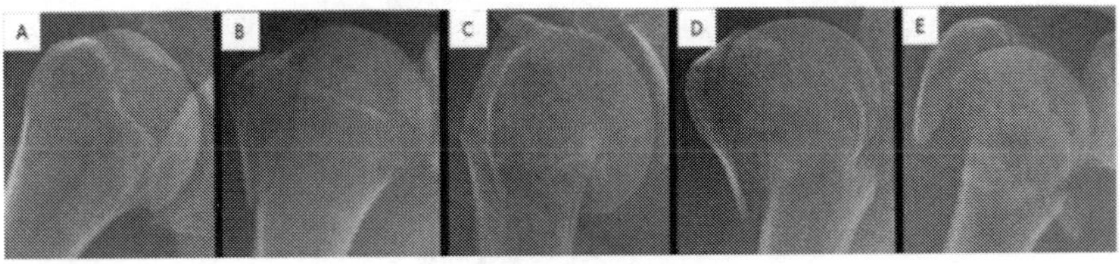

Figure 7. Neer's classification for proximal humerus [30].

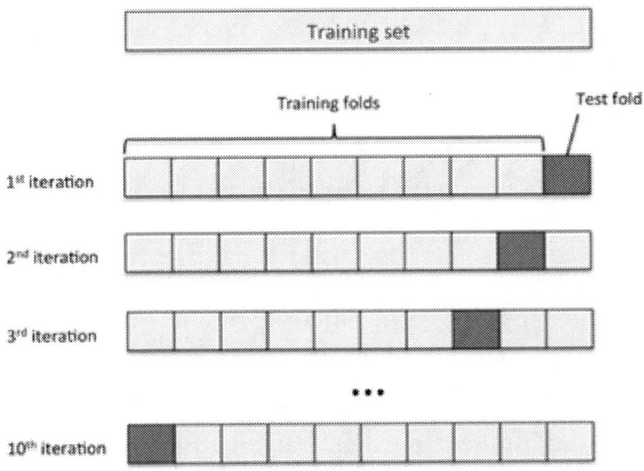

Figure 8. K-fold cross validation graphical representation. This process consist in dividing the dataset in K folds. For each iteration the network uses K-1 folds for training and the remaing fold for testing. In this way it is possible to obtain more generalized results.

Healthy humerus group is named A. Fracture classification was performed by two shoulder orthopedic specialists with 14 and 17 years of experience and one radiologist with expertise in musculoskeletal diseases and 15 years of experience. 515 cases were labelled as A, 346 cases as B, 514 cases as C, 269 cases as D, and 247 cases as E. The dataset of the 1.891 images was divided into 10 partitions without overlapping images: 1 partition was used as a test dataset, while all other images were used as training datasets. The authors used the open source pre-trained ResNet-152 [31] as a deep CNN model. As the dataset was divided into 10 partitions, 10 experiments were performed in order to obtain an averaged performance. The ResNet-152 showed superior performance to that of general physicians and general orthopedic surgeons and similar performance to that of the shoulder specialists. The authors also used 10-fold cross validation, showed in Figure 8, which is a good practice to reduce evaluation biases. The labelling phase should include different specialists, as this level of fracture is really complex to classify.

A limit of this work is the use of a specific classification for the proximal humerus fracture: to define a generalized approach, we should use a classification structure that can be applied to different bones. In the work of Jiménez-Sánchez et al. [32], the authors proposed a fully automatic CAD tool able to identify, localize and finally classify proximal femur fractures on X-Rays images according to the AO classification. Following AO classification, proximal femur fractures are divided into three main groups: A, B and C, depending from the area that is involved. Each of these classes are subsequently divided into sub-groups. The dataset is composed by a total of 1.347 X-Ray images. For the two classes problem 780 fracture images and 567 normal images were considered. The same dataset was used for the three class problem considering 327 images of type A fractures, 453 of type B fractures and 567 normal X-Rays. Three clinical experts participated in the evaluation: one 5th-year resident trauma surgeon, one trauma surgery attendant and one senior radiologist. Using AO foundation classification is a perfect approach to define a generalized method. The main obstacle broached by the authors is the large imbalance in the frequency of appearance of the classes in the fine-grained classification. A method to deal with unbalanced data must be defined.

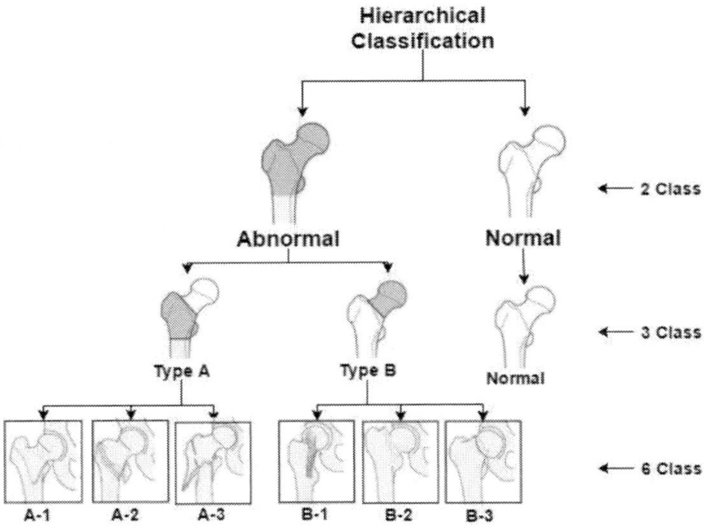

Figure 9. AO classification for the proximal femur case [32].

DISCUSSION AND CONCLUSION

Most of the work focused on classification between broken and unbroken bones, without extending the task to different types of fractures. We think that a generalized tool, able to distinguish different types of fractures, should follow the classification stated by the AO foundation. The AO classification is hierarchical, and is determined by the localization and configurations of the fracture lines, where each bone is divided in subsequent sub-groups of fracture, as shown in Figure 9 for the case of proximal femur. In the common literature, the AO classification was claimed to present a better reproducibility compared to other classification systems [33] and its configuration made it optimal for a classification task. Plus, the structure is the same for different bones in the human body, so the approach could be easily extended. Once defined the correct classification system, an adequate dataset is certainly one of the most important aspect for a Deep

Learning based application to operate efficiently. Even if good results were obtained for the fracture/no fracture classification without using a large dataset [14, 22] a correct number of images is suggested when the network has to distinguish different sub-groups of fractures. The dataset could be increased and balanced with data augmentation techniques, if needed, but without adding useless or misleading information. For example, using shear, strain or spot noise augmentation could cause a normal bone image to be classified as a bone with a fracture [22]. Thus, data augmentation is not always enough to balance irregular datasets. A solution could be to assign different weights to different classes when computing the cost function: classes with few images will be associated with higher weights. Also, technologies such as Generative Adversarial Networks [34] might be used to generate "fake" fractured bones, but this could probably be unfeasible as the fractures might result unrealistic. Concerning the pre-processing phase, the dataset should be cleaned from images containing prosthesis or other evident defects, and we recommend using no more than one image per person to decrease the over performance by the inclusion of a very similar image of the same patient. It is also demonstrated that selecting the fractured area and feed the neural network with cropped images instead of the full image improves the network performances [26, 32]. For this reason, a fully automated tool to select the fractured regions should be designed for the pre-processing phase. In addition, it is suggested to use lossless format such as PNG and TIFF and to resize images to different sizes and see which one works best [22]. Unfortunately, this is not feasible when using an existing network with transfer learning, because the input images must have a fixed size. Transfer learning allows to use a network, pre-trained on a different dataset, for your own dataset. The most used network architectures were VGG, ResNet, DenseNet and Inception, pre-trained with ImageNet dataset. As demonstrated in different papers, pre-training the network using a larger dataset of X-Ray bones images may improve the performance [18, 20]. For example, MURA dataset [20] is one of the biggest bones' dataset freely available. For this reason, we recommend to try different networks pre-trained with MURA dataset and test which one works best for the specific problem at hand. Cross-validation should be used to demonstrate that the network correctly generalizes the dataset features [30]. Another improvement could be introduced by removing surplus imaging data [30]. For example, if a network has to classify between the no fracture class and three different types of fractures A, B and C, it works better if trained excluding no fracture images. Following this results and the AO foundation classification, one idea should be to apply a hierarchical approach. To be more clear, a first network that classifies between fracture and no fracture and a sub-sequent one that takes the images predicted as fracture and categorizes them in A, B and C. Finally, Class Activation Mapping or similar technologies should be used to see where the network is focusing [20, 27, 32].

Table 1. Main strengths and limitations of each paper

	Main limitation	**Main strength**
Dimililer et al. [14]	Scant and bad-structured dataset	Pre-processing phase
Kim and MacKinnon [15]	Network pre-trained with a non-relatable dataset	Used transfer learning for the first time in this sub-field
Lindsey et al. [18]		Very large dataset for pre-training and training
Rajpurkar et al. [20]	Low results compared to the dimension of the dataset	The dataset used was made freely available by the authors
Olczak et al. [21]	Images labelled automatically without a second review	Used different networks structures
Yahalomi et al. [22]	Labelling carried on by just one specialist	Optimal results with a small dataset
Thurston et al. [26]	The region detection phase is semi-automated	Improved performance with the use of specific region
Cheng et al. [27]		Used CAM to visualize the learning process
Chung et al. [30]	Used a non-extendable classification method	Classification in different type of fractures
Jimenez-Sanchez et al. [32]	Imbalanced dataset	Used AO classification

Two papers do not present any main limitation: this does not mean that they do not have weaknesses, but that they are due to some missing aspects and not to some wrong implementations.

Last but not least, also the specialists have a fundamental role: the dataset must be correctly labelled, and different years of experience are needed to properly classify the types of fractures following the AO classification, especially the subgroups. Both for labelling and evaluation, if possible, more than one expert, coming from different specialization, should be enrolled in order to have multiple opinions. As the final aim of this tool would be to prove that the CAD system effectively help doctors in diagnosis, it would be also important to evaluate the performance of the specialists with and without the help of it [18]. A summary of the main limitations and strengths of each paper is showed in Table 1.

In this section we outlined the main aspects that should be taken in consideration when trying to tackle this task. This will be a first step in order to develop a solution that includes all these aspects and could be applied to classify all the bones in the human body, bringing a huge progress in the orthopedic field.

REFERENCES

[1] Amparore D, Checcucci E, Gribaudo M, Piazzolla P, Porpiglia F, Vezzetti E. Non-linear-Optimization Using SQP for 3D Deformable Prostate Model Pose Estimation in Minimally Invasive Surgery. In: Arai K, Kapoor S, editors. *Adv. Comput. Vis.*, vol. 943, Cham: Springer International Publishing; 2020, p. 477–96. https://doi.org/10.1007/978-3-030-17795-9_35.

[2] Olivetti EC, Nicotera S, Marcolin F, Vezzetti E, Sotong JPA, Zavattero E, et al. 3D Soft-Tissue Prediction Methodologies for Orthognathic Surgery—A Literature Review. *Appl Sci* 2019;9:4550. https://doi.org/10.3390/app9214550.

[3] International Osteoporosis Foundation. Broken bones, broken lives: a roadmap to solve the fragility fracture crisis in Europe 2018. https://www.iofbonehealth.org/broken-bones-broken-lives (accessed September 20, 2019).

[4] Krupinski EA, Berbaum KS, Caldwell RT, Schartz KM, Kim J. Long Radiology Workdays Reduce Detection and Accommodation Accuracy. *J Am Coll Radiol* 2010;7:698–704. https://doi.org/10.10160/j.jacr.2010.03.004.

[5] Hallas P, Ellingsen T. Errors in fracture diagnoses in the emergency department – characteristics of patients and diurnal variation. *BMC Emerg Med* 2006;6:4. https://doi.org/10.1186/1471-227X-6-4.

[6] Modhia U, Dickens A, Glezos C, Gehlert R, DeCoster T. Under-Utilization of the OTA Fracture Classification in the Orthopaedic Trauma Literature. *Iowa Orthop J* 2014;34:50–4.

[7] Checcucci E, Autorino R, Cacciamani GE, Amparore D, Cillis SD, Piana A, et al. Artificial intelligence and neural networks in urology: current clinical applications. *Minerva Urol Nefrol* 2019. https://doi.org/10.23736/S0393-2249.19.03613-0.

[8] Cirrincione G, Marcolin F, Spada S, Vezzetti E. Intelligent Quality Assessment of Geometrical Features for 3D Face Recognition. In: Esposito A, Faundez-Zanuy M, Morabito FC, Pasero E, editors. *Neural Adv. Process. Nonlinear Dyn.* Signals, vol. 102, Cham: Springer International Publishing; 2019, p. 153–64. https://doi.org/10.1007/978-3-319-95098-3_14.

[9] Kumar R. *Research Methodology: A Step-by-Step Guide for Beginners.* SAGE; 2019.

[10] Al-Ayyoub M, Hmeidi I, Rababah H. Detecting hand bone fractures in x-ray images. *JMPT* 2013;4:155–168.

[11] Cao Y, Wang H, Moradi M, Prasanna P, Syeda-Mahmood TF. Fracture detection in x-ray images through stacked random forests feature fusion. 2015 *IEEE 12th Int. Symp. Biomed.* Imaging ISBI, Brooklyn, NY, USA: *IEEE;* 2015, p. 801–5. https://doi.org/10.1109/ISBI.2015.7163993.

[12] Myint WW, Tun HM, Tun KS. Analysis on Detecting of Leg Bone Fracture from X-ray Images. *Int J Sci Res Publ IJSRP* 2018;8. https://doi.org/10.29322/IJSRP.8.9.2018.p8150.

[13] LeCun Y, Bengio Y, Hinton G. Deep learning. *Nature* 2015;521:436–44. https://doi.org/10.1038/nature14539.

[14] Dimililer K. IBFDS: Intelligent bone fracture detection system. *Procedia Comput Sci* 2017;120:260–7. https://doi.org/10.1016/j.procs.2017.11.237.

[15] Kim DH, MacKinnon T. Artificial intelligence in fracture detection: transfer learning from deep convolutional neural networks. *Clin Radiol* 2018;73:439–45. https://doi.org/10.1016/j.crad.2017.11.015.

[16] Szegedy C, Vanhoucke V, Ioffe S, Shlens J, Wojna Z. Rethinking the Inception Architecture for Computer Vision. 2016 *IEEE Conf. Comput. Vis. Pattern Recognit. CVPR*, Las Vegas, NV, USA: *IEEE;* 2016, p. 2818–26. https://doi.org/10.1109/CVPR.2016.308.

[17] Fei-Fei L, Deng J, Li K. ImageNet: Constructing a large-scale image database. J Vis 2010;9:1037–1037. https://doi.org/10.1167/9.8.1037.

[18] Lindsey R, Daluiski A, Chopra S, Lachapelle A, Mozer M, Sicular S, et al. Deep neural network improves fracture detection by clinicians. *Proc Natl Acad Sci* 2018;115:11591–6. https://doi.org/10.1073/pnas.1806905115.

[19] Ronneberger O, Fischer P, Brox T. U-Net: Convolutional Networks for Biomedical Image Segmentation. In: Navab N, Hornegger J, Wells WM, Frangi AF, editors. *Med. Image Comput. Comput.-Assist. Interv. – MICCAI* 2015, vol. 9351, Cham: Springer International Publishing; 2015, p. 234–41. https://doi.org/10.1007/978-3-319-24574-4_28.

[20] Rajpurkar P, Irvin J, Bagul A, Ding D, Duan T, Mehta H, et al. MURA: Large Dataset for Abnormality Detection in Musculoskeletal Radiographs. ArXiv171206957 Phys 2018.

[21] Olczak J, Fahlberg N, Maki A, Razavian AS, Jilert A, Stark A, et al. Artificial intelligence for analyzing orthopedic trauma radiographs: Deep learning algorithms—are they on par with humans for diagnosing fractures? *Acta Orthop* 2017;88:581–6. https://doi.org/10.1080/17453674.2017.1344459.

[22] Yahalomi E, Chernofsky M, Werman M. Detection of Distal Radius Fractures Trained by a Small Set of X-Ray Images and Faster R-CNN. In: Arai K, Bhatia R, Kapoor S, editors. *Intell. Comput.,* vol. 997, Cham: Springer International Publishing; 2019, p. 971–81. https://doi.org/10.1007/978-3-030-22871-2_69.

[23] Ren S, He K, Girshick R, Sun J. Faster R-CNN: Towards Real-Time Object Detection with Region Proposal Networks. *IEEE Trans Pattern Anal Mach Intell* 2017;39:1137–49. https://doi.org/10.1109/TPAMI.2016.2577031.

[24] Thurston M, MacKinnon T, Kim DH. Fracture detection with artificial intelligence: improved accuracy with region of interest focusing, 2018.

[25] Bradski G. The OpenCV Library. *Dr Dobb's Journal of Software Tools;* 2000.

[26] Thurston M, MacKinnon T, Kim DH. Fracture detection with artificial intelligence: improved accuracy with region of interest focusing, 2018.

[27] Cheng CT, Ho TY, Lee TY, Chang CC, Chou CC, Chen CC, et al. Application of a deep learning algorithm for detection and visualization of hip fractures on plain pelvic radiographs. *Eur Radiol* 2019;29:5469–77. https://doi.org/10.1007/s00330-019-06167-y.

[28] Selvaraju RR, Cogswell M, Das A, Vedantam R, Parikh D, Batra D. Grad-CAM: Visual Explanations from Deep Networks via Gradient-Based Localization. 2017 *IEEE Int. Conf. Comput. Vis*. ICCV, Venice: IEEE; 2017, p. 618–26. https://doi.org/10.1109/ICCV.2017.74.

[29] Huang G, Liu Z, Maaten L van der, Weinberger KQ. Densely Connected Convolutional Networks. 2017 *IEEE Conf. Comput. Vis. Pattern Recognit. CVPR,* Honolulu, HI: *IEEE;* 2017, p. 2261–9. https://doi.org/10.1109/CVPR.2017.243.

[30] Chung SW, Han SS, Lee JW, Oh K-S, Kim NR, Yoon JP, et al. Automated detection and classification of the proximal humerus fracture by using deep learning algorithm. *Acta Orthop* 2018;89:468–73. https://doi.org/10.1080/17453674.2018.1453714.

[31] He K, Zhang X, Ren S, Sun J. Deep Residual Learning for Image Recognition. 2016 IEEE Conf. Comput. Vis. Pattern Recognit. CVPR, Las Vegas, NV, USA: *IEEE;* 2016, p. 770–8. https://doi.org/10.1109/CVPR.2016.90.

[32] Jiménez-Sánchez A, Kazi A, Albarqouni S, Kirchhoff C, Biberthaler P, Navab N, et al. Towards an Interactive and Interpretable CAD System to Support Proximal Femur Fracture Classification. ArXiv190201338 Cs 2019.

[33] Jin WJ, Dai LY, Cui YM, Zhou Q, Jiang LS, Lu H. Reliability of classification systems for intertrochanteric fractures of the proximal femur in experienced orthopaedic surgeons. *Injury* 2005;36:858–61. https://doi.org/10.1016/j.injury.2005.02.005.

[34] Goodfellow IJ, Pouget-Abadie J, Mirza M, Xu B, Warde-Farley D, Ozair S, et al. *Generative Adversarial Networks.* ArXiv14062661 Cs Stat 2014.

In: An Introduction to Approaches … ISBN: 978-1-53618-680-2
Editor: Yi-Tung Chan

Chapter 4

DECISION-MAKING STRATEGIES WITH CLUSTERING BASED UNSUPERVISED LEARNING FOR SMART GRIDS PLANNING

Bogdan Constantin Neagu*, PhD
and Gheorghe Grigoraş, PhD
"Gheorghe Asachi" Technical University of Iasi,
Faculty of Electrical Engineering, Iasi, Romania

ABSTRACT

Technological developments in the electricity domain, including smart meters and smart grids, can provide unprecedented capabilities to the Distribution Networks Operators (DNOs) for shaping customer usage patterns, preventing outages, optimal unit commitment, etc. DNOs must be able to manage and use large databases and perform the advanced analytics necessary to make the best decisions. There are two main aspects of a decision-making system regarding the database management and data mining for analysis of data to gain certain knowledge or facilitate certain decision-making. Common scenarios to use unsupervised learning

* Corresponding Author's Email: bogdan.neagu@tuiasi.ro.

algorithms include data mining (DM), knowledge data discovery, outlier detection, or clustering process. These techniques help the Decison-Maker to understand the clumping structure of the data. These techniques help the Decision-Maker to understand the clumping structure of the data. The DNOs are alarmed by the variable behavior of power consumption and the outliers from the recordings regarding the irregularity, undiscovered power demand, energy thefts, or inadequately installed devices. The chapter presents a clustering-based decision-making approach using an unsupervised learning data mining process that acquires the knowledge discovery in databases (KDD) techniques, focused on the identification of outliers from the records provides by Smart Meters (SMs). By applying this approach, the DNOs could know with high accuracy the real technical losses and establish proper operation and planning strategies of actually "smart" networks. The case studies that use the real distribution networks of a Romanian DNO highlighted the robustness of the approach.

Keywords: strategies, clustering techiques, smart grids, data mining, deep learning, load profiling, forecasting, power losses, planning

ACRONYMS

Artificial Intelligence – AI
Cophenetic Correlation – CC
Data Mining – DM
Distribution Network Operator – DNO
European Union – EU
High voltage – HV
Knowledge Data Discovery – KDD
Low Voltage – LV
Medium Voltage – MV
Silhouette Voefficient – SC
Smart Meters – SMs
Smart Metering System – SMS
Typical Load Profile – SVV
Smart Grids – SGs

1. Introduction

The required demand of the consumers from the nodes of the electrical distribution networks has a variation that depends on time and location of consumption. Thus, the DNOs need information about the consumers' demand to be used in the optimal planning and operation of the networks, identification of the optimal supply solutions, load management, and the billing with high accuracy. The required demands of consumers depend on several factors, such that [1, 2]:

- *the consumer' class*. It is identified through the consumption category, electric heating type (with/without), the size of the building, and so on;
- *the time*. It is influenced by the hour, day type (work or weekend), month, and season [3];
- *the climatic and meteorological factors*. They refer to the temperature, humidity, cloudiness, wind speed, and so on;
- *the correlation between the analysed demand and other demands*;
- *the records, as the load profiles, from a certain period* [4].

For a particular consumer, his behaviour is represented by a load profile that depends on the energy consumption for a certain period (day, week, month, or year), and consumption category. Generally, the consumers with low demand (mainly residential ones) have a poorly described behaviour, due to the implementation of the smart metering system (SMS) at this level, in a short period, lead to high investments from the side of DNOs, with long recovery time [5, 7]. For these consumers, there is only information about electricity consumption for certain periods (for example, every two months in some countries, including Romania) [6]. For consumers from other categories, as is the case for industrial consumers, the installation of smart meters is facilitated by the advantages related to billing and high electricity consumption. The billing is done every month, and the investment is justified by a detailed record of consumption, allowing the application of differentiated tariffs according to the consumption period.

Regarding the records of load curves, they can be associated with the individual consumer curves or the aggregation process at the level of the low voltage (LV) or medium voltage (MV) feeder, the MV/LV electric distribution substation, or the High Voltage/Medium Voltage (HV/MV) electric station [7].

The classic approach uses in the planning of the electric distribution systems the load profiles to estimate the maximum load with the help of a simultaneous coefficient of the consumers fed from an MV/LV electric distribution substation. Although this approach can be considered as appropriate, there are several main disadvantages in its use [2, 6]:

- the peak load is only known;
- inherent inaccuracies appear due to the simultaneity coefficients which must be calculated;
- the calculation of the energy consumption, respectively of the losses has a small accuracy;
- the values of voltage in all nodes of the network are not known in each moment;
- the nodes with random variations of the demand do not have the load profiles modelled with high accuracy.

The new methods, based on data mining, can model the load profiles with more high accuracy [8, 9, 10, 11]. These can be afterward in the calculation of power and energy losses for certain periods (day, week, month, or year). The main advantages of the load profiles refer to the following aspects [2, 6]:

- the calculation or estimation of the simultaneous coefficients does not be made because the load profiles include this information;
- the steady-state regime is calculated more accurate, and the voltages, power flows, loadings, and power/energy losses on each branch (line or transformer) are known for any analysis interval;

- the optimal control intervals of the taps for each transformer can be evaluated;
- the load forecasting is more accurate than in the classical approach.

The LV and MV electrical distribution networks, especially urban ones, have very more nodes, even in the case when only the distribution substations are considered. MV consumers (mainly the industrial consumers) are also very different and have power demands which vary inside of large intervals, between 10 kVA and 1000 kVA. In these conditions, the monitoring of the consumption in each node is very difficult and seldom is practically impossible to be achieved. But this issue can be removed by the DNO, if they use the typical load profiles, assigned to each node, in the steady-state regime calculations of the own electrical distribution networks where the monitoring is missing [6].

2. CLUSTERING IN DATA MINING TO KNOWLEDGE EXTRACTION

2.1. General Aspects

Clustering represents a special process to group the data $[X]_{nxm}$, represented by *n* elements introduced as vectors $[X_i]_{1xm}$, $i = 1, \ldots,$ n, based on the spatial distribution of their *m* characteristics. To study the similarity/difference between the elements associated wih a set of data with the purpose grouping them, each element will be defined through a vector $[X_i]$, $i = 1, \ldots, n$, with *m* characteristics/ attributes [6].

$$[X]_{nxm} = \begin{bmatrix} X_1 \\ X_2 \\ \cdots \\ X_n \end{bmatrix}_{nx1} = \begin{bmatrix} x_{11} & x_{12} & \cdots & x_{1m} \\ x_{21} & x_{22} & \cdots & x_{2m} \\ \cdots & \cdots & \cdots & \cdots \\ x_{n1} & x_{n2} & \cdots & x_{nm} \end{bmatrix}_{nxm} \quad (1)$$

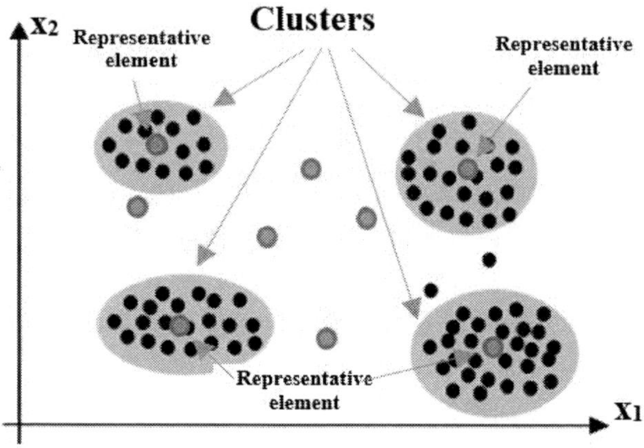

Figure 1. The clusters clustering process.

The number of characteristics and their definition establish based on a deeper analysis of the database and the experience of the Decision-Makers. The vectors $[X_i]$, $i = 1, \ldots, n$, with m characteristics, associated with the analysed database, are merged into different clusters, according to the distance calculated between each of them. Finally, one or more clusters (groups/patterns/models/classes) will be obtained in function by the spatial location of the characteristics of vectors $[X_i]$, $i = 1,\ldots, n$ [12, 13].

The characteristics of vectors will be closer to the common centre of the cluster they belong, compared to centres of other clusters. This aspect is exemplified in Figure 1, where the vectors $[X_i]$, with two characteristics (x_{i1} and x_{i2}), are represented using circles in different colours.

The black circles represent the vectors assigned to a cluster, the blue circles were used for the unassigned vectors, and the red colour corresponds to the centres of clusters. The vectors are grouped using a distance-based similarity criterion. Each cluster is characterized by a representative element, with the mean characteristics of all vectors belonging to the respective cluster. The unassigned vectors represent so-called outliers. Their identification must be made in the initial pre-processing stage, such that the cluster formation process to be not affected [6].

2.2. Stages of the Clustering Process

The stages of the clustering process are shown in Figure 2 [6].

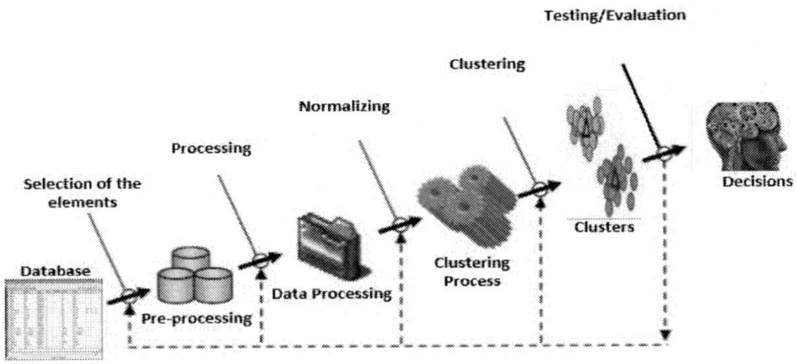

Figure 2. The stages of the clustering process

Selection of the elements. The elements which fulfill the following objects are chosen from the database, identifying, also, the type and number of characteristics.

Extracting the characteristics of the elements. It refers to the identification of the most useful and representative characteristics of the elements, subject to the clustering process. During the extraction process, to obtain new dominant characteristics, one or more element transformations can be performed.

Normalization process of values. The values of characteristics are normalized using a suitable normalizing factor. This process is it is mandatory because all data must be brought in the range [0, 1] to be grouped.

Defining a similarity measure. A distance establishes the similarity between elements, defined as vectors. Many measurement systems of the distance are defined, but from them, the Euclidean distance-based system is the most used. Thus, the distance can be calculated, in the case of two vectors X_r and X_s, with the relation [12, 13]:

$$d(X^r, X^s) = \sqrt{(X_r - X_s)(X_r - X_s)^t} \quad (2)$$

The clustering process. The clustering process can be carried out in several ways, depending on the methods chosen by the decision-maker. All clustering methods should lead to obtaining the clusters, indifferently by the input data. There are cases when the representative clusters could not be obtained. Another method can be applied to obtain better results than previous cases to identify a solution. The quality of the results can be the best when the separation of the elements leads to the well-defined clusters or less good when a membership degree to one of the clusters is associated.

Extraction of results. The representation of data must be made in a simple form, easy to interpret by the Decision-Maker to interpret the results. Simplicity is needed either from two perspectives: automatic analysis, with the help of a computing system that can perform efficiently the subsequent data processing, or the human when the representation of data is easier to be understood by Decision-Makers. In the clustering process, extracting the results is a brief description of each cluster, using an average value of the characteristics associated with the representative elements.

Testing/Evaluation of results. The validity of the obtained results, represented by the clusters, is based on an analysis of the clustering process and usually uses an optimization criterion. This analysis is objective and determines whether the results obtained are correct. The validation of clusters is made if these do not appear "accidental" or for other reasons [6, 14–17].

2.3. Clustering Methods

There are several ways to classify the clustering methods. The most of them use the classification is two categories [6, 12, 13]:

- hierarchical methods;
- K-means method.

Hierarchical methods can be classified into agglomeration and division methods, in function by the sense of clustering progress. In the case of agglomeration methods, it starts from the *K* clusters, each with a single element, and finally, a single cluster is obtained through successive mergers. The division methods consider an opposite sense of clustering progress to the hierarchical methods, such that *K* clusters are finally obtained, each with a single element [X_i], starting from a single cluster. Agglomeration techniques are more used.

As previously shown, in the hierarchical clustering process, a clustering process undergoes a series of mergers. The first step consists of *n* clusters, with a single element, and the last step contains a single cluster having all elements. At each stage, the method mergers two close clusters. At the first level, this means the merger of two elements that are close to each other (as a distance), because each cluster has only one element. Starting from the above description, for most commonly used hierarchical clustering methods are given the equations on which they are based.

Single linkage method

$$\mathrm{d}(\mathrm{C_r},\, \mathrm{C_s}) = \min_{\substack{X_r \in C_r \\ X_s \in C_s}} \{d(X_r, X_s)\} \tag{3}$$

where $d(X_r, X_s)$ represents the Euclidean distance.

Complete linkage method

$$\mathrm{d}(\mathrm{C_r},\, \mathrm{C_s}) = \max_{\substack{X_r \in C_r \\ X_s \in C_s}} \{d(X_r, X_s)\} \tag{4}$$

Average linkage method

$$\mathrm{d}(\mathrm{C_r},\, \mathrm{C_s}) = \frac{\sum\limits_{\substack{X_r \in C_r \\ X_s \in C_s}} d(X_r, X_s)}{|N_r| \cdot |N_s|} \tag{5}$$

where N_r and N_s represents the number of elements from the clusters C_r and C_s, respectively.

Centroid method

$$d(C_r, C_s) = \left\| \overline{X_r} - \overline{X_s} \right\|^2 \tag{6}$$

where $\overline{X_r}$ and $\overline{X_s}$ are the average vectors of the clusters C_r and C_s.

Ward Method

$$d(C_r, C_s) = \frac{\left\| \overline{X_r} - \overline{X_s} \right\|^2}{\frac{1}{N_r} + \frac{1}{N_s}} \tag{7}$$

K-means method

$$J = \min \left\{ \sum_{k=1}^{K} \sum_{l=1}^{n_k} \left\| X_l^k - c_k \right\|^2 \right\} \tag{8}$$

where $\left\| x_l^k - c_k \right\|^2$ represents the distance between the vector $X_l{}^k$ and the centroid of cluster c_k, $k = 1, ..., K$, $l = 1,..., n_k$; n_k represents the number of vectors from the cluster k.

2.4. The Validation of Clustering Results

The evaluation of the quality is very important to validate the clusters. The characteristics examined are density, size, shape, separation, and the robustness of the classification. There are more tests to validate the results of the clustering process [6, 14–18].

- *External Tests*. The results of the classification are compared with data from a testing database.
- *Internal Tests*. The application of these tests is done to each cluster to be validated.

- *Relative Tests*. These tests consider more classifications of the input data. The final results are compared with others, obtained with the same method, but with data from a testing database.

After linking the objects in a data set into a hierarchical cluster tree, it might want to verify that the distances (that is, heights) in the tree reflect the original distances accurately. Also, it might want to investigate natural divisions that exist among links between objects. In a hierarchical cluster tree, any two objects in the original data set are eventually linked together at some level. The height of the link represents the distance between the two clusters that contain those two objects. This height is known as the cophenetic distance between the two objects. If the clustering is valid, the linking of objects in the cluster tree should have a strong correlation with the distances between objects in the distance vector.

The closer the value of the cophenetic correlation coefficient is to 1, the more accurately the clustering solution reflects your data [17].

$$CC = \frac{\sum\left(d(X_r,X_s) - \overline{d(X_r,X_s)}\right)\left(d(T_r,T_s) - \overline{t(T_r,T_s)}\right)}{\sqrt{\left(\sum\left(d(X_r,X_s) - \overline{d(X_r,X_s)}\right)^2\right)\left(\sum\left(t(T_r,T_s) - \overline{t(T_r,T_s)}\right)^2\right)}} \tag{9}$$

where:

$d(X_r, X_s)$ - the Euclidean distance between vectors $[X_r]$ and $[X_s]$, $r \neq s$ from the database [X].

$t(X_r, X_s)$ - the dendogrammatic distance between the models $[T_r]$ and $[T_s]$, $r \neq s$ from the dendogram [T].

$\overline{d(X_r,X_s)}$, $\overline{t(X_r,X_s)}$ - the average distances.

Another technique calculates the widths of the silhouette for each sample, each cluster, and the total data set. Thus, a so-called silhouette characterizes each cluster based on the comparison of its tightness and separation. The width of the average silhouette is finally used to evaluate the quality of the clustering process [16].

$$SC = \frac{1}{K}\sum_{c=1}^{K}\left(\frac{1}{N_c}\sum_{l=1}^{N_c}\left(\frac{d_{e_l^c} - h_{e_l^k}}{\max\{d_l, h_l\}}\right)\right) \tag{10}$$

where

$d_{e_l^c}$ – average distance between element e_l^c, ($l = 1, ..., n_c$) and objects of the same cluster c;

$d_{e_l^k}$ - minimum average distance between the element e_l^c, ($l = 1, ..., n_c$) and elements from the closest cluster k to cluster c ($k \neq c$).

A value of the SC coefficient closer to 1 indicates that a strong structure has been found.

3. The Load Profiling in the Electric Distribution Networks

3.1. The Load Profiling Process

In this section, a clustering-approach to determine the typical load profiles for the nodes from the electric distribution networks, with the 20 kV voltage level, is presented.

The number of nodes from the analysed network is 34. The typical load profiles-based approach represents an alternative to the smart metering-based approach. Knowing the load profile of the nodes, the DNOs can simplify the demand forecasting in that supply area. Thus, they can provide better efficiency in marketing strategies.

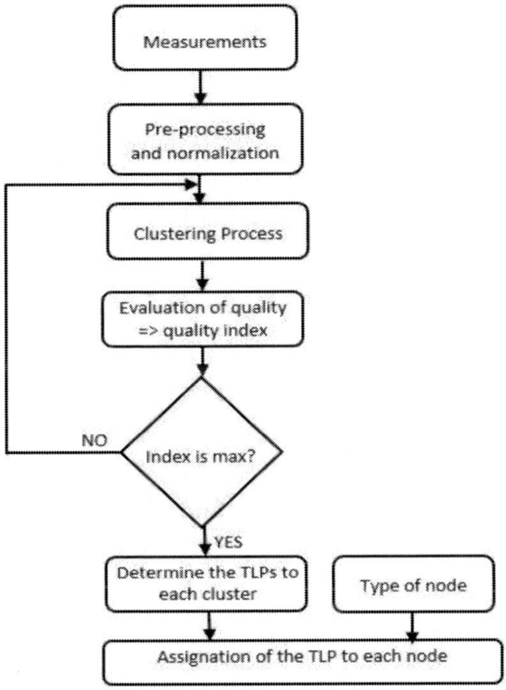

Figure 3. The flowchart of the clustering-based profiling process.

The load profile of each node is reconstructed using the normalized load profile and their daily (monthly, yearly, depending on the case) energy consumption.

The sampling interval of the load is one hour, such that the typical load profile is represented by 24 values throughout the day. The database is presented in form of matrix $[P]_{nxm}$, where *n* represents the number of nodes (in this analysis, *n* is equal with 34) and *m* is equal with 24 (hours of a day).

Due to a large number of load profiles associated with various nodes that can create additional problems in their handling, the proposed approach merges the load curves into clusters based on the similarities between them. A typical load profile, TLP, is determined for each cluster. For each MV node is assigned such a profile, using the steps depicted in Figure 3 [18].

$$[P]_{34x24} = \begin{bmatrix} P_1 \\ P_2 \\ \cdots \\ P_{34} \end{bmatrix} = \begin{bmatrix} p_{1,1} & p_{1,2} & \cdots & p_{1,24} \\ p_{2,1} & p_{2,2} & \cdots & p_{2,24} \\ \cdots & \cdots & \cdots & \cdots \\ p_{34,1} & p_{34,2} & \cdots & p_{34,24} \end{bmatrix} \quad (11)$$

3.2. Data Pre-Processing Stage

3.2.1. The Analisys of Big Data

The Smart Grids is an electrical network that uses digital technology to monitor and control the electricity consumption between generation to end-users. Following the requirements of the European Union (EU), 80% of end-users must-have devices withy smart measures before 2020 [19]. In many data analysis assignments, a large number of variables are being recorded or sampled. One of the first steps towards obtaining a coherent analysis is the detection of outlying observations. Although outliers are often considered as an error or noise, they may carry important information. Detected outliers are candidates for aberrant data that may otherwise adversely lead to model misspecification, biased parameter estimation and incorrect results [20].

The analysis of data represents the starting point for many applications, in the design or operation phase for online control or complex processes. Nowadays, the need to extend the capabilities of human analysis for handling the overwhelming quantity of data that we can collect has become increasingly necessary. Since computers have enabled the possibility of storing large amounts of data, it is only natural to resort to computational techniques to help us discover meaningful patterns and massive structures volume data [21].

The load profile plays a fundamental role in operation and planning of power systems. Unfortunately, due to various random factors, the load curves always contain abnormal, deviation, unrepresentative, noisy, strange, anomalous, and missing data. It is fundamental for power systems operators to detect and repair anomalous or abnormal data before the use of a load curve in planning and modeling process.

For this reason, a deep learning-based data mining approach for the detection of outliers using the two stages is proposed. In the first step, knowledge discovery on database technique for extraction of the load curves' main indicators, computed with the information provided by Smart Meters, was used. The second stage coincides with the detection of irregular consumption, outliers, from the above-computed indicators, using a statistic-based data mining technique. The proposed method was tested on a real database with 300 electric substations from the rural areas. The results highlighted the ability of the proposed approach to be very efficient in the decision-making process regarding the operation and planning of power systems using only useful load curve characteristic information provided by a large database through data mining techniques.

However, the biggest danger of outliers is its impact on statistical results, which could lead to misleading conclusions. In the business world, decision making, based on misleading conclusions, can be very costly and devastating for a DNO. Therefore, outliers must be firstly detected and, subsequently, decide on how to treat them. Outliers might be just mistyped values, but they may also point to important factors in business processes [22].

The online monitoring of the consumers' loads supposes the use of a large number of meters from SMS, spread over a large geographical area and a long period, such that, there is the possibility as, in some situations, different technical issues to affect the quality of the records from the database. The most relevant and frequent issues refer to the interruptions of the communication support, the failure of the meters, and the atypical or irregular behavior of certain consumers.

The final result of these issues reflects in the records from the database that contain null, missing values, or exceeding a certain threshold established by the connection approval. Once these data are removed or treated with the missing data techniques, the database can be used in the clustering process.

3.2.2. Outlier Detection Using a Statistical Deep Learning Based Data Mining Approach

The ability of Decision-Maker to analyze an un-automated large database is far overrun. Because the memory of the computer allows recording more information, the abnormal data can be detected using some proper algorithms. Only Artificial Intelligence (AI) or statistical methods can be used in the analysis of large databases. The record that deviates from the most common information was considered as an outlier. Consequently, these outliers are the most drawback in electricity values of the daily load curve. For outliers' detection, two methods can be used, mainly, data mining techniques and steady-state estimation [23]. At the same time, the references from the literature propose other data mining approaches based on the artificial neural network [24], fuzzy [25], clustering algorithms [26], and time series methods [27].

Data mining can lead to obtaining useful information unknown from measurements [28]. At the same time, data mining extracts characteristics [29] from the generated databases by the software tools [30]. The steps for the KDD process are described in [31]. The useful step in the KDD process is represented by Data mining between the transformed data and patterns. Data mining can fit the determination of the pattern using recordings from the database. Notwithstanding, the knowing tools used in the data mining process are the following [32]: clustering methods, classification analysis, association laws, approximation algorithms, subordination, and forecast analysis.

The data mining approaches were split up into two categories, which contain supervised and unsupervised learning algorithms. In the first category, an operational model is obtained only for the modeled variable. The model displays the typical variable with another of them. The classification and regression methods are usually considered from this category [33].

The necessary information is extracted from the database using data mining techniques. After the outliers are identified and removed using the statistics-based approach, it follows the calculation of the main indicators of load curves. The proposed methodology is adapted to the Gaussian distribution with an implicit assumption that uses information characterized by the first two statistical times, as is described in Figure 4. The probability density from these cases coincides with the real data estimation using a simple approach computation. A dissonant outlier in a database consists of a data differently from the rest of them, believed to be generated by a new technique regarding other data [34].

A very large number of measurements can generate Gaussian distribution with a perfect balance if the mean is exactly in the middle of the interval. The curve has a specific bell shape that might be wider, more spread out, or narrower, closer to the mean. Often, this distribution occurs in each data measurement.

Mathematically, taking into account that μ is the average value, one standard deviation represented as $\mu \pm \sigma$ is 68% of entire data, and two standard deviations have up to 95% of the entire data, see Figure 4, [35].

The outside values represent the outliers. For a total sampling number (T_s), the standard deviation (σ) is:

$$\sigma = \frac{1}{T_s}\int_0^{T_s} X^2(t)dt - \left(\frac{1}{T_s}\int_0^{T_s} X(t)dt\right)^2 \tag{12}$$

and the average value:

$$\mu = \frac{1}{T_s}\int_0^{T_s} X(t)dt \tag{13}$$

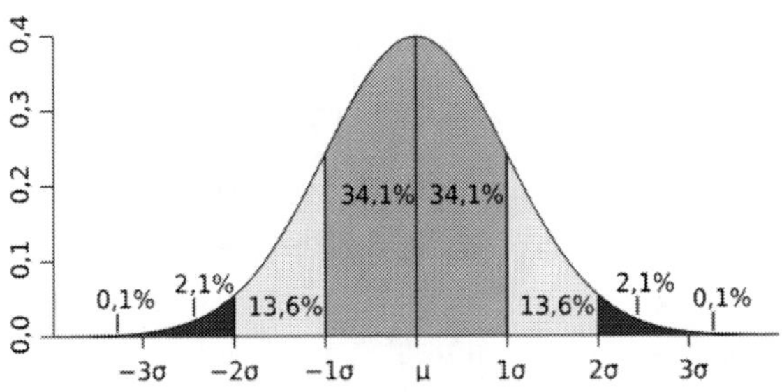

Figure 4. The Gaussian distribution.

The outliers (noted as Λ) can be rewritten using the above statistic indicators from (12) and (13), as:

$$\Lambda = \mu \pm \Psi \cdot \sigma \tag{14}$$

where Ψ is 2 for this approach, being evaluated after a deeper analysis of all information provided from the Smart Meters by use another statistical based data mining criterion [34].

3.3. Normalization

All records are normalized considering a suitable factor regarding the average power, peak power, or energy consumption over the surveyed period, as:

$$p_{ij}^{n} = \frac{p_{ij}}{\sum_i p_{ij}} \quad i = 1, \ldots, 34; j = 1, \ldots, 24 \tag{15}$$

where: p^{n}_{ij} - normalized value of the hourly power j, required the node i;
p_{ij} – the hourly power j, required the node i;
$\sum_i p_{ij}$ - the energy over the surveyed period (24 hours).

3.4. Clustering Process

Because the classification of the load profiles through the visual comparison of the graphics is subjective and impractical, the cluster analysis method was applied to solve this problem. The main objective of the clustering analysis is to compare elements, represented by vectors associated with the load profiles, and to gather them progressively in the coherent groups, such that the profiles in the same group to be similar. Differences between methods arise due to the different criteria in defining the distance between clusters.

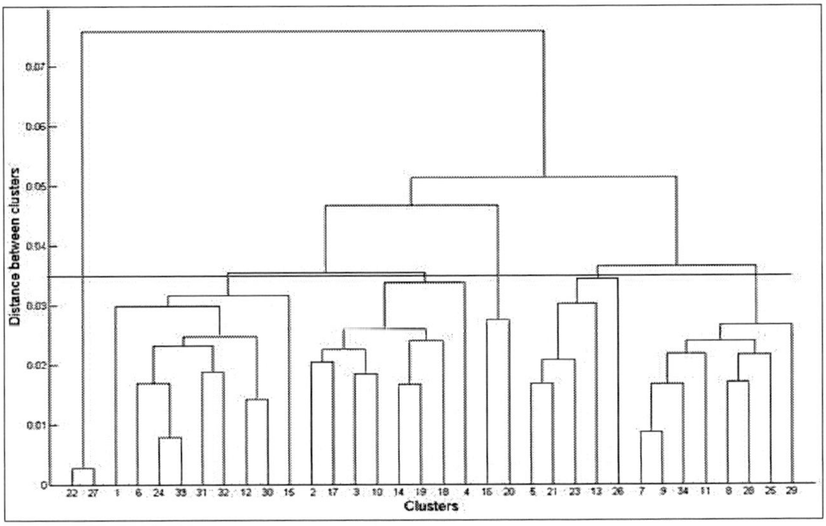

Figure 5. The dendogram of Average Linkage method.

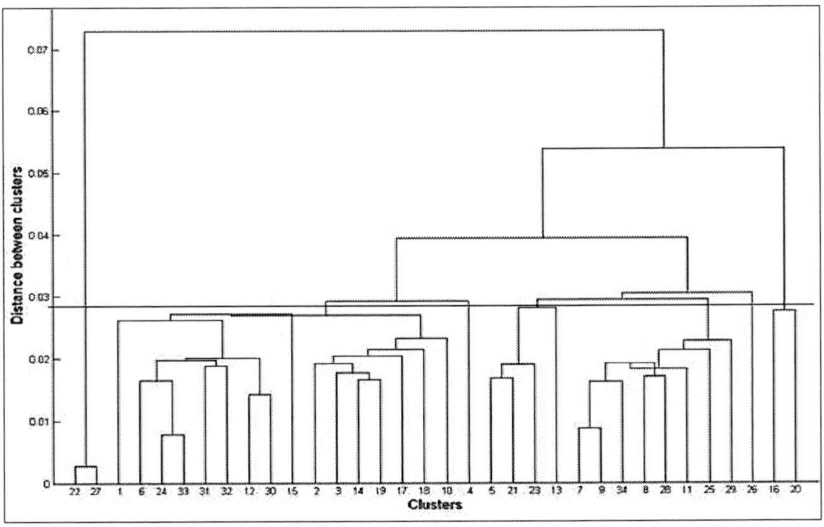

Figure 6. The dendogram of Centroid method.

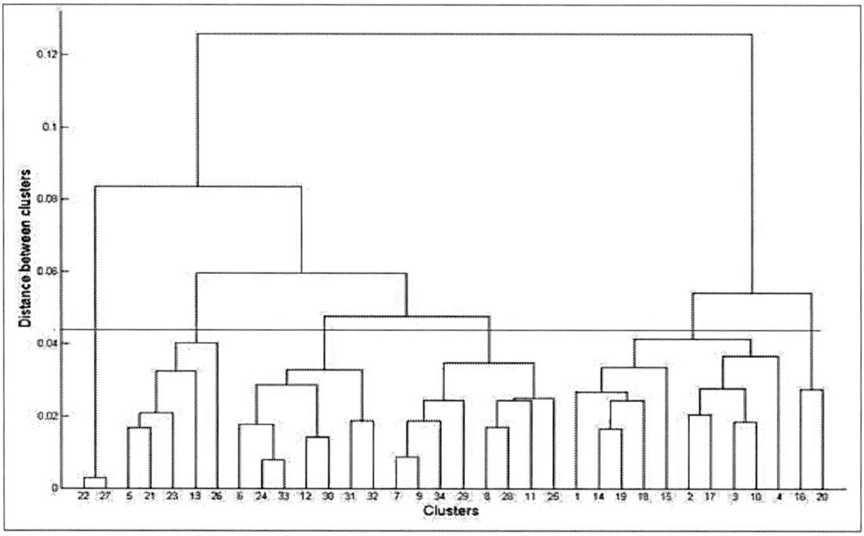

Figure 7. The dendogram of Complete Linkage method.

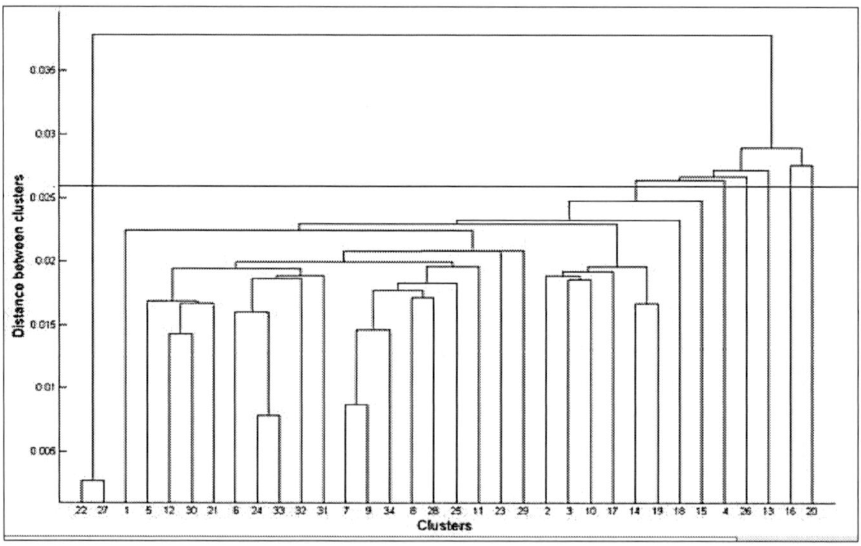

Figure 8. The dendogram of Single Linkage method.

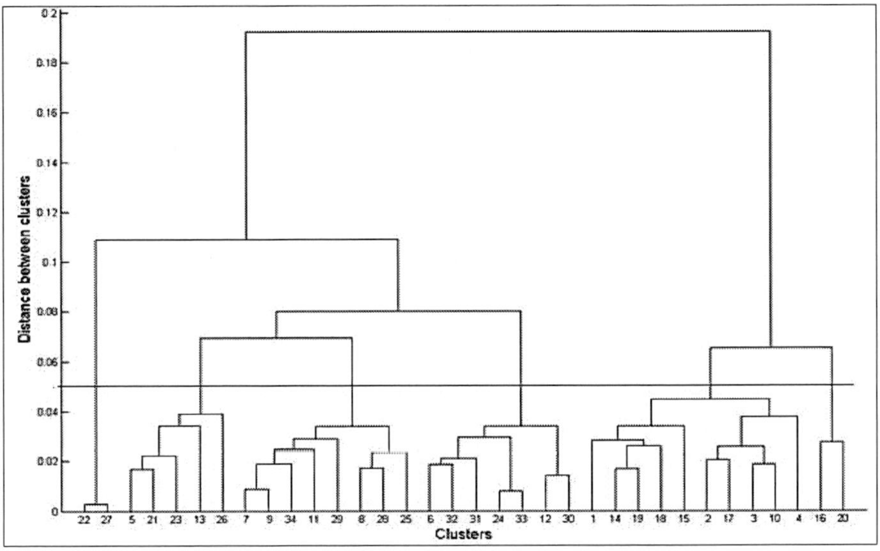

Figure 9. The dendogram of Ward method.

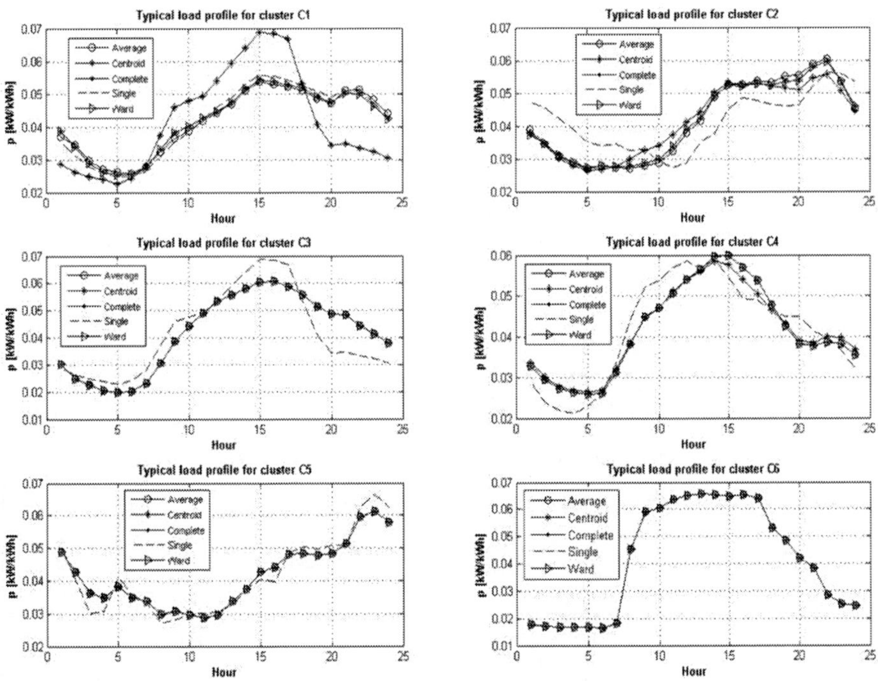

Figure 10. The comparison between the TLPs obtained with hierarchical methods.

The results of the clustering process, using five methods, are presented, as dendrograms, in Figures 5-9. The height of the blue line indicates the distance between clusters. As can be seen from figures, the first difference between dendrograms refers to the range of the distances between clusters. The red line indicates the threshold imposed in each method to obtain the clusters.

The final results of the clustering process, identified by TLP, are shown in Figure 10. The analysis of the results highlights differences between methods, the most obvious is in the case of the Centroid and Single Linkage methods. The clusters C1 and C2 contain more load profiles than in the case of other methods. The quality of the clustering process is highlighted through two coefficients: cophenetic correlation (CC) and silhouette (SC) in Table 1.

Table 1. The clusters and quality coefficients for different hierarchical methods

Methods	C1	C2	C3	C4	C5	C6	CC coefficient	SC coefficient
Average Linkage	9	8	8	5	2	2	0.743	0.45
Centroid	1	17	8	4	2	2	0.739	0.43
Complete Linkage	7	10	8	5	2	2	0.683	0.41
Single Linkage	28	1	1	1	1	2	0.600	0.12
Ward	7	10	8	5	2	2	0.740	0.44

It can be observed that the most efficient method is the average linkage method, which has the highest values of the CC and SC coefficients. Similar performances were obtained in the case of the Ward method. It can be concluded that the methods influence the final results, with significant differences from a database to another.

4. Energy Losses Estimation for Smart Grids Planning Based on Deep Learning Methods

4.1. Smart Grids Planning. Introduction

The concepts of Smart Grids (SGs) defined in both the industrial and academic environments, take different forms, focusing attention on several particular issues regarding the active consumers, distributed generation, or active participation in the electricity market.

Data from the smart meters allow detailed analyses on the operation and planning of electric networks, giving a strategic advantage to the DNOs to identify the areas with a small performance. In the SGs, the benefits are win-win between the actors (DNO, consumers, and energy renewable producers). On the other hand, the currents on the three phases should have equal values to operate in balancing symmetric regime. But, due to the unequal distribution of the consumers amongst the three phases along with variations, in their demand appear the unequal loading of phases, the so-called "current unbalance" [36]. In this context, SGs planning is an essential step in the design and operation processes. To accentuate this aspect, another two case studies are presented in the following. The first study presents the deep learning-based energy losses forecasting, which is tested in a SG with 100% from meters integrated into the SMS. In the second case, the influence of outliers on the transformer power loss estimation, using the proposed statistical-based data mining methodology, is presented.

4.2. The Active Losses Forecasting Using a Deep Learning-Based Expert System

This paragraph deals with a major interest issue to electricity suppliers and DNOs: energy losses assessment in low voltage (LV) electrical networks. Currently, electrical distribution networks supply a growing number of single-phase consumers, with unbalanced phase distribution, which determines inefficient operating conditions for these networks.

The traditional approach computes the energy losses using two specific coefficients called the "*time of losses*" and the "*maximum load.*" The drawback of this approach refers to the use of the subjective value of these coefficients and significant errors in the assessment maximum apparent power. Thus, to assess and forecast the active energy losses in the LV distribution networks, an efficient methodology based on the real measurements provided by smart meters and using unsupervised learning methods, is described.

The proposed methodology uses a typical load profile estimation using a deep learning approach based on a proper expert system for the detection of outlier using the load information provided by SMS. The deep learning method consists of an expert system to detect the outliers having the process described in Figure 11.

A real LV rural distribution network from the East of Romania is used to highlight the accuracy of the proposed methodology. The proposed methodology processes all end-users' energy load profiles extracted directly from the Data Warehouse (SMS database). The load profiles are shown in Figure 12.

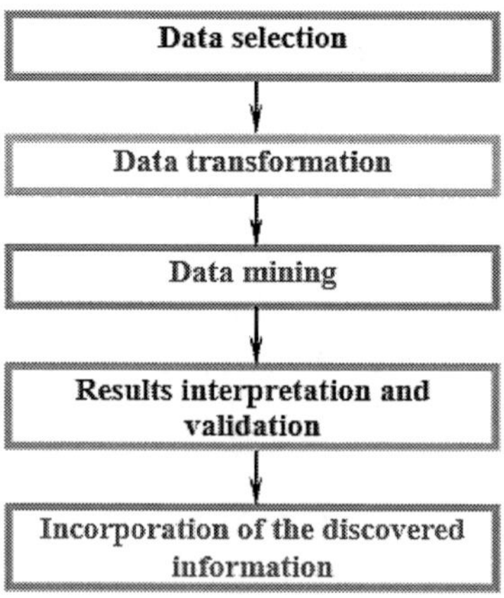

Figure 11. Expert system for outlier detection using a deep learning data mining method.

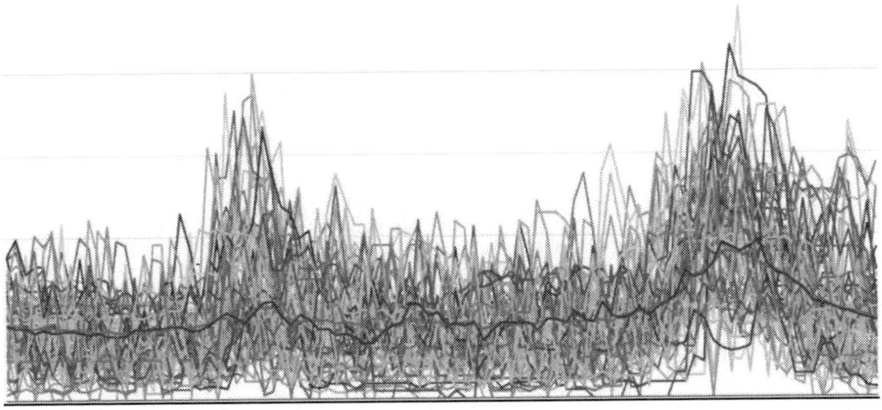

Figure 12. Real energy load profilc to all cnd-uscrs.

In the processing step the typical load profile process begin by computing the average value of active power as a particular form of eq. (15) from the normalization step, with:

$$P^* = \frac{1}{n}\sum_{k=1}^{t} P_k^* = \frac{1}{n}\sum_{k=1}^{t} \frac{P_k}{W_k} \tag{16}$$

where: P_k and W_k is the active power and energy of the consumer k;

n is the maximum number of consumers;

t is number of sampling from the analysed period.

In the processing step, the profiling process begins by computing the average value of active power as a particular form of eq. (15) from the normalization step, with:

$$P^* = \frac{1}{n}\sum_{k=1}^{t} P_k^* = \frac{1}{n}\sum_{k=1}^{t} \frac{P_k}{W_k} \tag{16}$$

where: P_k and W_k is the active power and energy of the consumer k;

n is the maximum number of consumers;

t is number of sampling from the analysed period.

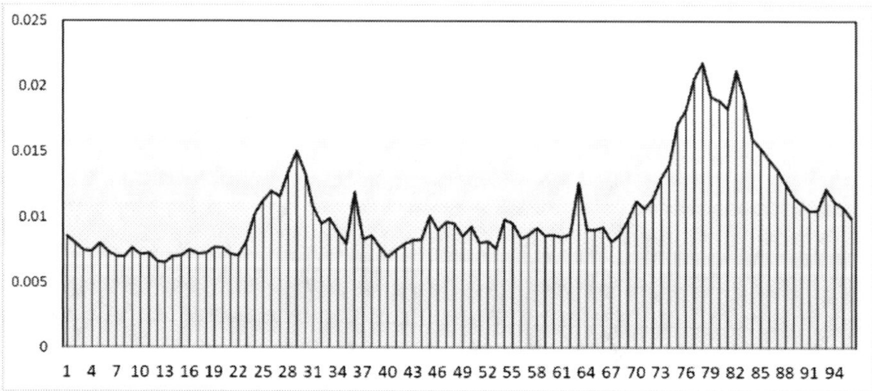

Figure 13. The typical load profile resulted using our proposed approach.

Using the data mining step (Figure 11), all curves with consumption equal with 0 were neglected, and the energy load profile forecasted for the all consumers presented in Figure 2 uses the following expression:

$$P_{med_i} = P^* \cdot W_k \tag{17}$$

where P_{med} is the average value of estimated active power.

The estimated typical load profile used subsequently for the energy losses forecasting is presented in Figure 13. Must mention that the sampling step is 15 minutes. In the abscissa, 96 samples of 15 minutes corresponding to a 24-hourly day can be identified.

The active energy losses from the distribution networks can be obtained using iterative load flow computation for a necessary and sufficient number of levels [34]. For the considered distribution network, the mathematical model leads to the solving of nonlinear algebraic equations, such as the following:

$$\begin{aligned} P_i &= G_{ii} \cdot U_i^2 + \sum_{\substack{k=1 \\ k \neq i}}^{n} U_i \cdot U_k \left[G_{ik} \cdot \cos(\delta_i - \delta_k) + B_{ik} \cdot \sin(\delta_i - \delta_k) \right], \quad i \neq e \\ Q_i &= -B_{ii} \cdot U_i^2 - \sum_{\substack{k=1 \\ k \neq i}}^{n} U_i \cdot U_k \left[B_{ik} \cdot \cos(\delta_i - \delta_k) - G_{ik} \cdot \sin(\delta_i - \delta_k) \right], \quad i \neq e \end{aligned} \tag{18}$$

where: P_i, Q_i – active and reactive powers injected in the network node i;

G_{ii}, G_{ik}, B_{ii}, B_{ik} – real and imaginary elements of the nodal admittance matrix;

U_i, U_k, δ_i, δ_k – modules and arguments of voltages in analysed network nodes i and k.

By adopting some additional simplifying assumptions (considering the models of elements with reactance, estimating the voltages in the nodes through their rated values and eliminating the trigonometric functions, taking into account that the phase differences of the voltages at the terminals of the elements are low), the equations system (18) becomes linear, such as:

$$P_i = U_{n,i} \sum_{\substack{k=1 \\ k \neq i}}^{n} U_{n,k} \cdot B_{ik} \cdot (\delta_i - \delta_k); \quad i = \overline{1,n}; i \neq e \tag{19}$$

where $U_{n,i}$, $U_{n,k}$ – nominal voltages of the i and k nodes.

By solving the equations system (19) the active power flows on all network elements are:

$$\begin{aligned} P_{ik} &= B_{ik} \cdot U_{n,i} \cdot U_{n,k} \cdot (\delta_i - \delta_k) \\ P_{ki} &= -P_{ki}; \quad i = \overline{1,n}; \qquad i \neq k \end{aligned} \tag{20}$$

and active power losses on the entire power network shall be determined as:

$$\Delta P = \sum_{i=1}^{n-1} \sum_{k=i+1}^{n} \frac{P_{ik}^2}{U_{n,i} \cdot U_{n,k}} R_{ik} \tag{21}$$

This determinist method using for the active energy losses evaluation through repeated load flow computation, the daily active/reactive load curves with a sufficient number of load levels were considered. The daily active energy losses are computed with the relation:

$$\Delta W_{day} = \sum_{k=1}^{N_p} \Delta P(k) \cdot \Delta t_k \tag{22}$$

where N_p is the number of levels from daily load curves and Δ_{tk} the duration of the k level from the load curves.

Results. The case study was conducted on a rural network. The proposed network has a total length of 3 km, suppling 172 single-phases households (109 on the first feeder and 63 on the second feeder) and the public lighting. The on-line diagram is presented in Figure 14. The line length between 2 posts was considered as 40 m, following Romanian standards.

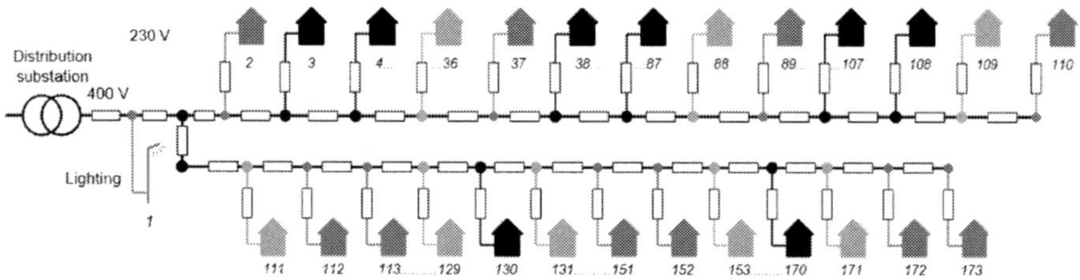

Figure 14. Single-line diagram of the LV distribution network.

The energy losses in the unbalance LV network were computed using iterative load flow computation in DigSilent Power Factory application, based on the Newton-Raphson load flow algorithm [34]. As input data, the measured three-phase active and reactive power and the single-phase end-users' locations (phase A, B, or C) were used. However, input data are necessary to specify the distribution network topology and, also, the impedances and lengths of each overhead line branch.

For instance, to assess the energy losses in the LV distribution network, the simulation was conducted for a day, with sampling steps of 15 and 60 minutes, which means 96 and 24 load flow calculations. Secondly, the energy losses were forecasted by using the typical load profile forecasted (from Figure 13), keeping the above assumptions. The summary of results is given in Table 2. To validate the energy losses method, the obtained values are reported to the total active energy injected in the electric substation (582.95 kWh).

The energy losses are indicated in Table 3 only for the first branches of each feeder F1 and F2 (Figure 14), for the unbalanced case. It can be observed that the proposed forecasting method (the end-users consumption profiled with a characteristic load profile) led to a smaller error.

Table 2. Summary results for the daily energy losses

Sampling step	Input data	Active energy losses	
		[kWh]	[%]
15 minutes	Real	100.84	17.3
	Forecasted	100.46	17.2
1 hour	Real	102.23	17.5
	Forecasted	97.00	16.6

Table 3. Summary of energy losses for unbalanced state for firsts branch F1 and F2

Imput data	Phase	15 minutes		1 hour		15 minutes	1 hour
		F1 [kWh]	F2 [kWh]	F1 [kWh]	F2 [kWh]	F1+F2 [kWh]	F1+F2 [kWh
real	A	2.23	0.24	2.2	0.24	2.47	2.44
	B	1.51	0.73	1.6	0.72	2.24	2.32
	C	0.51	0.17	0.52	0.16	0.68	0.68
	N	0.54	0.28	0.56	0.24	0.82	0.8
	total	4.78	1.42	4.88	1.36	6.2	6.24
Forecasted	A	0.24	1.49	0.24	1.48	1.73	1.72
	B	0.71	2.38	0.72	2.28	3.09	3
	C	0.16	0.51	0.16	0.52	0.67	0.68
	N	0.20	0.49	0.2	0.48	0.69	0.68
	total	1.31	4.88	1.28	4.72	6.19	6

Also, the errors associated with the active energy losses, assessed with real information provided from smart meters and the typical load profile, are presented in Figure 15. Regarding the unbalanced steady-state regime, the errors between the energy losses estimation with different sampling steps, using both forecasting approaches, were indicated in Figure 16 for the two first branches of the LV distribution network proposed for analysis.

In the two presented case studies, the average value of the energy losses is 1.81% for real, respectively 1.61 for our approach. Also, it can be seen that the error is a maximum of 5.2% in the case of the 1-hour sampling step, and for the 15-minutes sampling step, the proposed methodology has smaller errors, between 0.2% and 0.4%.

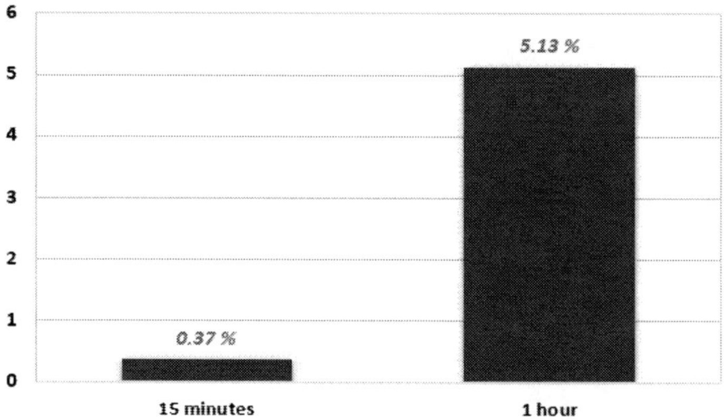

Figure 15. The errors between the two methods, for both smapling steps.

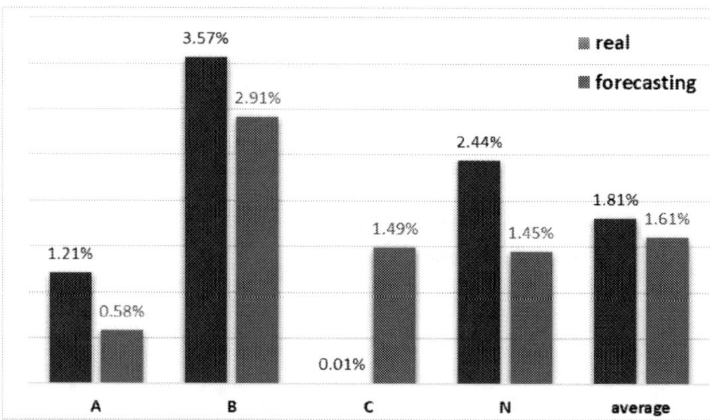

Figure 16. The errors between the two methods, for both sampling steps, on the first branch of the two LV network feeders.

Thus, the proposed approach can be implemented with success in assessing and forecasting the active energy losses for all LV distribution networks, including the classic meters.

4.3. The Influence of Outliers on Active Losses Forecasting on Power Transformers Level

The power transformers technical losses comprise two types of losses the so-called no-load or iron losses ($\Delta P_{no\text{-}load}$) and load or cooper energy losses (ΔP_{load}). Then, the power losses in each power transformer can be estimated as:

$$\Delta W_T = \Delta P_{no\text{-}load} \cdot T + \Delta P_{load} \cdot \left(\frac{P_{max} + Q_{max}}{S_n} \right)^2 \cdot \tau \quad (23)$$

where: S_n represent the nominal power of transformers;

T are maximum load duration;

P_{max}, Q_{max} are the maximum values of active and reactive powers that flows through the transformers;

τ – loss duration.

In this paragraph, a real SMS database, containing the active and reactive load curves provided by 300 power transformers from the MV/LV substations, was explored. All substations are equipped with Smart Meters, installed on the LV side, to record the active and reactive powers, voltages, power factors on single-phase and three-phases. The measurements correspond to one week with a 10-minutes sampling step.

Table 4. The Primary data of power transformers from distribution substations

Type	Number of transformers	Nominal Power [kVA]	Active losses		Transformer Resistance [Ω]
			No-load [kW]	Load [kW]	
1	9	40	0.230	1.0	0.100
2	47	63	0.300	1.5	0.060
3	84	100	0.350	2.3	0.037
4	92	160	0.525	3.1	0.019
5	56	250	0.680	4.4	0.011
6	9	400	0.940	6.0	0.006
7	1	630	1.250	8.2	0.003

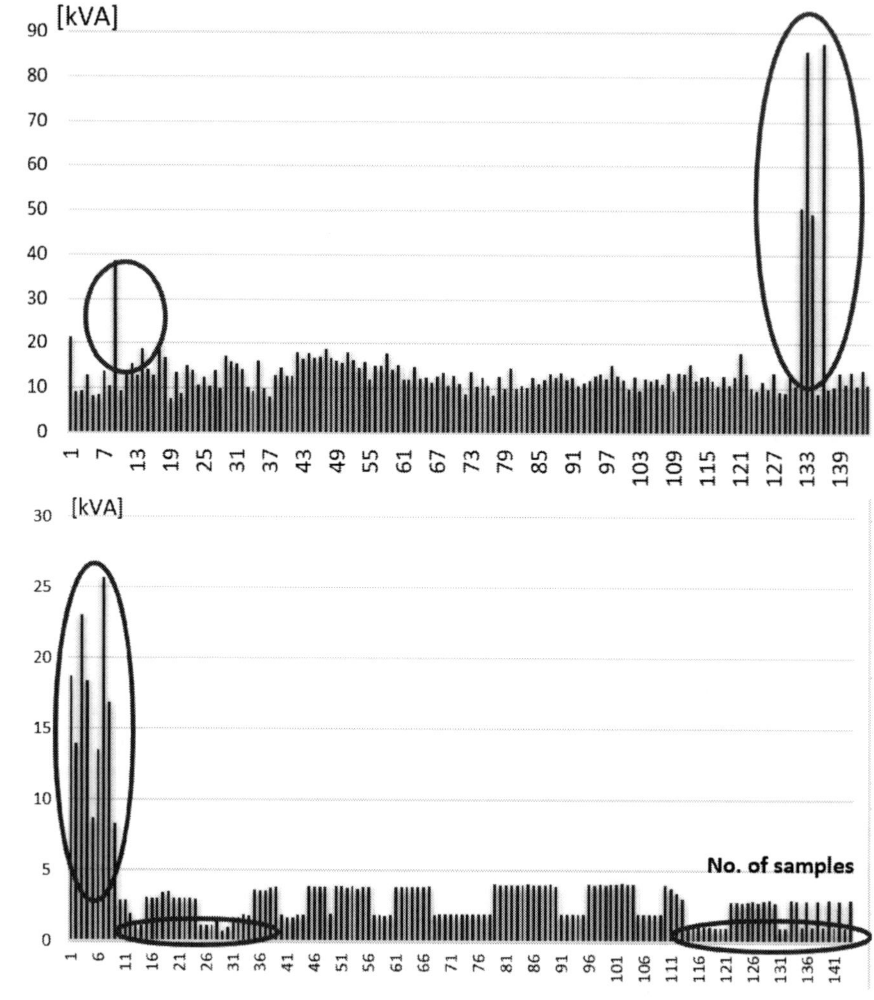

Figure 17. Load curve with outliers provided by Smart Meters.

For all considered transformers, the rated powers, resistances, no-load losses, and load losses, are presented in Table 4.

In Figure 17 are presented the apparent power curves (with 10-minutes sampling step), in the case of two substations, where the red circles highlight the outliers.

In the first step, by using the mathematical relations already proposed in [35], the three indicators defined in (23), were calculated: load factor (Smax = Pmax+jQmax), the peak load value of the load duration, loss duration (τ), and maximum load duration (T).

Next, the indicators of the load curves, for every substation, were computed applying the proposed data mining process, using eq. (12) ÷ (14).

To validate the obtained values for the three indicators, the Pearson coefficients, defined as a measure of the linear correlation between two variables, are applied, see Table 5.

The values of loss duration and the maximum load duration (between the dotted lines) for a characteristic day (Wednesday), and the outliers (outside the dotted lines), computed with (14), are presented in Figure 18.

Table 5. The Pearson coefficients

Indicators	**S_{max}**	**T**	**τ**
S_{max}	1.000	0.985	0.899
T	0.985	1.000	0.887
τ	0.899	0.887	1.000

It follows a data-cleaning process, and the resulted values of outliers are indicated in Table 6 for the two indicators.

Table 6. The detection of outliers from considered load curve indicators

Load Indicators	**Outliers, *A***				
	min	***max***	***No.***	μ	σ
T, hours	3.142	18.729	12	10.936	3.897
τ, hours	0.578	13.908	13	6.665	3.621

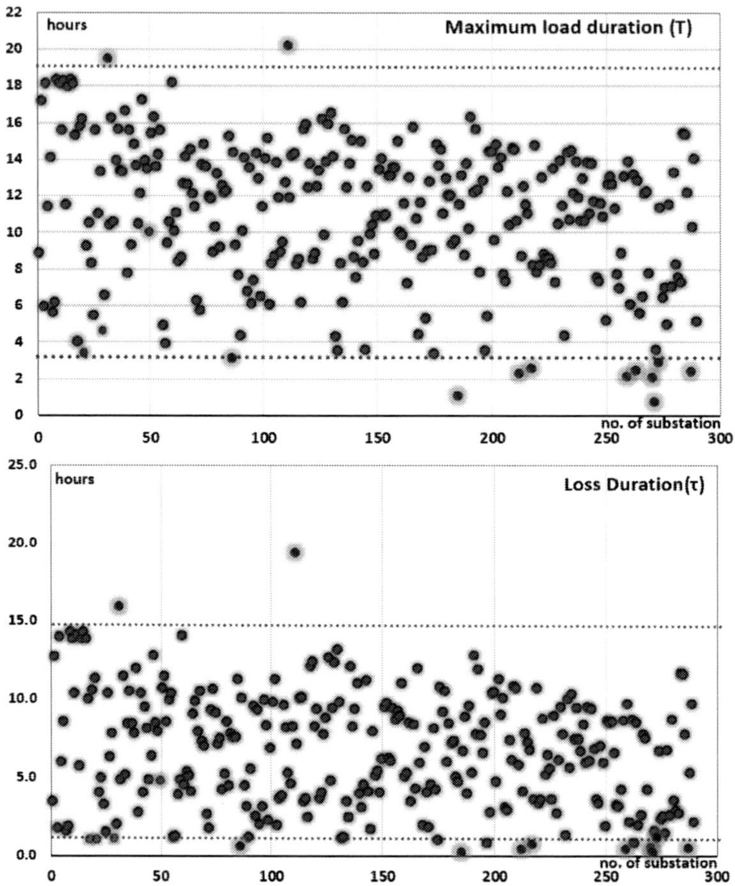

Figure 18. The maximum load duration and loss duration for the two substations.

At this time, ten power substations were removed from our analysis, such that in the final point, only the 290 substations were considered. Using the proposed approach, finally, we can characterize the power stations from the processed curves. The analyses were made on the transformer groups (rated power and numbers).

The database was divided into seven transformers categories: 40 kVA (9), 63 kVA (45), 100 kVA (80), 160 kVA (90), 250 kVA (56), 400 kVA (9), and 630 kVA (1). For the accuracy of values, in Table 7, the results (without outlier) for the seven transformers' categories are exposed.

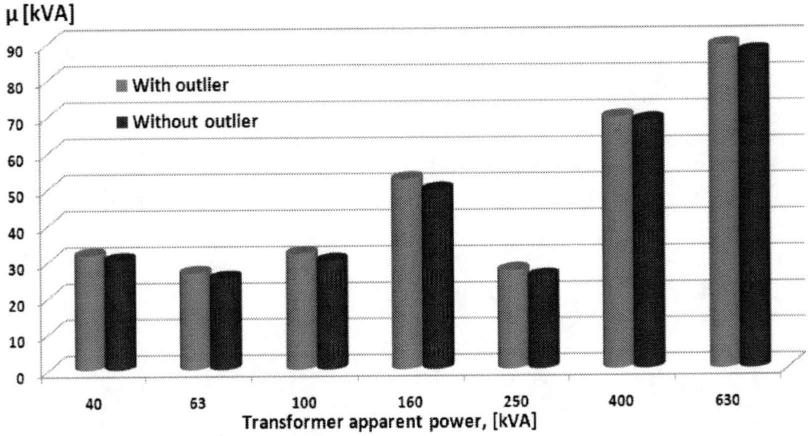

Figure 19. The average apparent power for the all transformer with and without outliers

Table 7. The statistical results of the load curves indicators for each transformers category

Indicators		T	τ	Indicators		T	τ
40 kVA	μ	0.674	0.519	160 kVA	μ	0.625	0.489
	σ	0.059	0.055		σ	0.113	0.110
63 kVA	μ	0.626	0.491	250 kVA	μ	0.627	0.493
	σ	0.102	0.101		σ	0.090	0.080
100 kVA	μ	0.615	0.483	400 kVA	μ	0.663	0.505
	σ	0.098	0.097		σ	0.060	0.064
				630 kVA	μ	0.720	0.552

Furthermore, the obtained maximum value of the apparent power flow through all considered transformer categories, with and without outliers, is shown in Figure 19. These values are founded in the energy losses estimation computation.

The technical losses are presented in Figure 20, where the influence of the outliers on the active energy losses was also indicated. The total losses are computed using (23), for all power transformer categories.

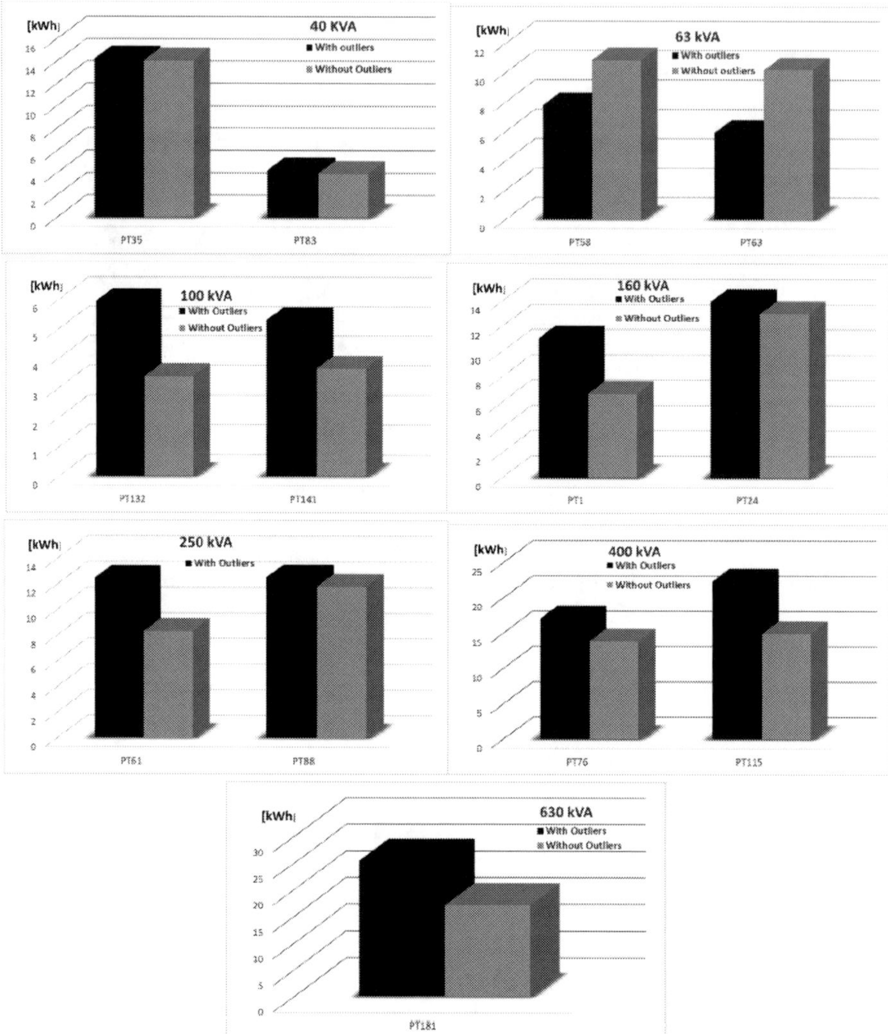

Figure 20. Active power losses for all transformer categories.

CONCLUSION

The planning and operation of smart grids require efficient management of large databases, which is essential in the decision-making process, influencing optimal supervision and control. Thus, it is necessary to improve the routine calculations, with emphasis on data classification and the implementation of complex tasks regarding selective analysis, multiobjective optimization, or optimal on-line control.

In this context, the practical issues related to optimal supervision and control can be solved, combining classical numerical techniques with the facilities offered by data mining algorithms. These approaches are used in all stages of the decision-making process, but especially to overcome the difficulties from the multi-objective modeling, handling of constraints, uncertainty considerations like load estimation, costs, or loads.

In the chapter, a clustering-approach to determine the typical load profiles for 34 nodes from the electric distribution networks, with the 20 kV voltage level, was proposed. Special attention was paid to the data pre-processing, where the data mining techniques were used to identify the outliers, which can affect the final results.

In the clustering process, the hierarchical methods were used, and their influence on the final results was analysed. The analysis of quality indexes (cophenetic correlation and silhouette coefficients) highlighted the differences between methods.

The most obvious was in the case of the Centroid and Single Linkage methods. The best results were obtained after the application of the Average Linkage and Ward methods. Not all methods lead to the best results regarding the TLPs assigned the nodes from the electric distribution networks. Therefore, to identify the method with the best partitions of the load profiles, which influences obtaining the TLPs, a deeper analysis must be done.

Also, the results obtained highlight that outliers can be easily identified and then eliminated. Afterward, energy losses can be calculated more accurately. It observed that there was a significant decrease in active energy losses if the outliers are eliminated.

If the TLPs are known, the DNOs can simplify the demand or power losses forecasting in the supply areas with the similar characteristics of the consumers. Thus, they can provide better efficiency in planning and operation strategies.

References

[1] Classification of electricity load forecasting based on the factors influencing the load consumption and methods used: An-overview, (2015) *IEEE Conference on Energy Conversion* (CENCON).

[2] Carter-Brown, C. G., (1999), Load profile modeling for integrated energy planning, *Domestic Use of Electrical Energy Conference*, pp. 13–18.

[3] Khuntia, S. R., Rueda, J. L., and van der Meijden, M. A. M. M., (2016) Forecasting the load of electrical power systems in mid- and long-term horizons: a review, *IET Generation, Transmission & Distribution, 10, 16*, pp. 3971 – 3977.

[4] Tao, S., Li, Y., Xiao, X., and Yao, L., (2017), *Load forecasting based on short-term correlation clustering, Innovative Smart Grid Technologies - Asia (ISGT-Asia)*, Auckland, New Zealand, pp. 1-7.

[5] Imani, M., and Ghassemian, H., (2018), Electrical Load Forecasting Using Customers Clustering and Smart Meters in Internet of Things, *9th International Symposium on Telecommunications* (IST), Tehran, Iran, pp. 1-7.

[6] Grigoraş, G., Scarlatache, Fl., and Neagu, B. C., (2016), *Clustering in Power Systems. Applications*, Lambert Academic Publishing, Germany.

[7] Madhavi, K. L., Cordova, J., Ulak, M. B., Ohlsen, M., Ozguven, E. E., Arghandeh, R., and Kocatepe, A., (2017), *Advanced electricity load forecasting combining electricity and transportation network, North American Power Symposium (NAPS)*, Morgantown, WV, USA, pp. 1-6.

[8] Din, G. M. U., and Marnerides, A. K., (2017), Short term power load forecasting using deep neural networks, *International Conference on Computing, Networking and Communications (ICNC)*, Santa Clara, CA, USA, pp. 594 – 598.

[9] Hanjie, L., Huajun, Z., Qinqin, H., Yu, S., and Fuwen, S., (2018), Power Load Forecasting Based on Data Analysis and Neural

Networks, *2018 Chinese Automation Congress (CAC)*, Xi'an, China, pp. 744-748.

[10] Fahiman, F., Erfani, S. M., Rajasegarar, S., Palaniswami, M., and Leckie, C., (2017). Improving load forecasting based on deep learning and K-shape clustering, *International Joint Conference on Neural Networks (IJCNN)*, pp. 4134-4141.

[11] Shahzadeh, A., Khosravi, A., and Nahavandi, S., (2015), Improving load forecast accuracy by clustering consumers using smart meter data, *International Joint Conference on Neural Networks (IJCNN)*, Killarney, Ireland, pp. 1-7.

[12] Gordon, A. D., (1999), *Classification*, 2nd ed., Chapman & Hall, NewYork, USA.

[13] Jain, A. K., Murty, M. N., and Flynn, P. J., *Data Clustering: A Review*, [Online]. Available: www.cermics.enpc.fr/~keriven/vision/ articles.

[14] Razaee, M. R., Lelieveldt, B. P. F., and Reiber, J. H. C., (1998), A New Cluster Validity Index for the Fuzzy C-means, *Pattern Recognition Letters, vol. 19*, pp. 237 - 246.

[15] Yatskiv, I., and Gusarova, L., (2005), The Methods of Cluster Analysis Results Validation, *Proc. of International Conference RelStat'04*, Riga, pp. 75 – 80.

[16] Rousseeuw, P. J., (1987), Silhouettes: a Graphical Aid to the Interpretation and Validation of Cluster Analysis, *Journal of Computational Applied Mathematics*, vol. 20, pp. 53 – 65.

[17] Saraçli, S., Doğan, N., and Doğan, İ., (2013), Comparison of hierarchical cluster analysis methods by cophenetic correlation, *Journal of Inequalities and Applications*, Article 203, available on-line: https://journalofinequalitiesandapplications.springeropen.com /articles/10.1186/1029-242X-2013-203.

[18] Grigoraș, G., Barbulescu, C., Cârțină, G., and Comanescu, D., (2011), A Comparative Study Regarding Efficiency of the Hierarchical Clustering Techniques in Typical Load Profiles Determination, *Acta Electrotehnica*, vol. 52, no. 5, pp. 184 – 188.

[19] Levenda, A., Mahmoudi, D., and Sussman, G., (2015), The Neoliberal Politics of "Smart": Electricity Consumption, Household Monitoring,

and the Enterprise Form," *Canadian Journal of Communication*, v. 40, n. 4, pp. 615-636.

[20] Queiroz, L., Cavellucci, C., and Lyra, C., (2009), Evaluation of technical losses estimation in LV power distribution systems. In *CIRED 2009-20th International Conference and Exhibition on Electricity Distribution*, Prague, Czech Republic.

[21] Fidalgo, J. N., Matos, M., and Marques, M., *Estimation and Allocation of Losses in Distribution Networks*, available on-line: http://www.aedie.org/9CHLIE-paper-send/350-FIDALGO.pdf.

[22] Neagu, B. C., and Grigoras, G., (2017), Detection of Irregular Consumption to Load Monitoring in Smart Grids Environment," *22nd IMEKOTC4 International Symposium & 20th International Workshop on ADC Modelling and Testing*, Iasi, Romania.

[23] Diksha, M., and Raghavendra, T. S., (2017), A Survey on Big Data Energy Based on Smart Grid, *Int. J. of Adv. Res., Id. and Innov. in Tech.*, 3(1), pp. 594-598.

[24] Rakhshani, E., Sariri, I., and Rouzbehi, K., (2009), Application of data mining on fault detection and prediction in boiler of power plant using artificial neural network, *International Conference on Power Engineering, Energy and Electrical Drives*, pp. 473-478.

[25] Cateni, S., Colla, V., and Vannucci, M., (2007), A fuzzy logic-based method for outliers detection, in *Artif. Intellig. and App*, pp. 605–610.

[26] Loureiro, A., Torgo, L., and Soares, C., (2004), Outlier Detection using Clustering Methods: a Data Cleaning Application, in *Proc. of KDNet Symp. on Knowledge-based Syst. for the Public Sector*, Bonn, Germany.

[27] Tang, G., Wu, K., Lei, J., Bi, Z., and Tang, J., (2014), From landscape to portrait: A new approach for outlier detection in load curve data, *IEEE Trans. Smart Grid*, 5(4), pp. 1764-1773.

[28] Abiteboul, S., Arenas, M., Barceló, P., and Bienvenu, M., (2016), Research directions for Principles of Data Management, *SIGMOD Record,* 45(4), pp. 5–17.

[29] Jia, N., Wang, J. S., and Li, N., (2012), Application of data mining in intelligent power consumption, *Int. Conf. on Autom. Control and Artifi. Intelligence*, pp. 538-541.

[30] Neagu, B. C., and Grigoras, G., (2015), Data Mininig Tools in Electricity Distribution Systems, *Acta Electr. J.*, 56(3), pp. 209-212.

[31] Gilchrist, M., Mooers, D., Skrubbeltrang, G., and Vachon, F., (2010), Knowledge discovery in databases for competitive advantage, *Journal of Management & Strategy*, 3(2), pp. 2-15.

[32] Neagu, B. C., Grigoras, G., and Scarlatache, F., (2017), Effects of outliers on calculation of load profile factors. *International Conference on Modern Power Systems* (MPS), pp. 1-4.

[33] Neagu, B. C., Grigoras, G., and Scarlatache, F., (2018), Influence of Outliers on Transformer Power Losses Estimation Using a Statistical Based Data Mining Approach, *10th International Conference on Electronics, Computers and Artificial Intelligence (ECAI),* Iasi, Romania, pp. 1-4.

[34] Neagu, B. C., Grigoras, G., and Scarlatache, F., (2017), Outliers discovery from Smart Meters data using a statistical based data mining approach, *10th International Symposium on Advanced Topics in Electrical Engineering* (ATEE), Bucharest, pp. 555-558.

[35] Neagu, B. C., Grigoras, G., and Ivanov, O., (2019), *The Optimal Operation of Active Distribution Networks with Smart Systems*. In Advanced Communication and Control Methods for Future Smartgrids. IntechOpen, 2019, pp. 1-24.

[36] Neagu, B. C., Grigoras, G., Scarlatache, F., Schreiner, C., and Ciobanu R. C., (2017), Patterns discovery of load curves characteristics using clustering based data mining," *11th IEEE International Conference on Compatibility, Power Electronics and Power Engineering (CPE-POWERENG),* Cadiz, Spain, pp. 83-87.

In: An Introduction to Approaches … ISBN: 978-1-53618-680-2
Editor: Yi-Tung Chan

Chapter 5

USE OF ENSEMBLE LEARNING TECHNIQUES TO ANALYZE DATA RELATED TO EDUCATION, HEALTH AND STANDARD OF LIVING

Juhi Gajwani* and Pinaki Chakraborty, PhD
Department of Computer Science and Engineering,
Netaji Subhas University of Technology, New Delhi, India

ABSTRACT

In this chapter, we will present our analysis of the Subnational Human Development Index (SHDI) data of the states (provinces) within India. It has three components, namely, health, education and standard of living. This index provides data individually about regions within a country and can be used to examine the disparities that exist therein. We experimented with different factors that might, in some way, influence any of the three dimensions of SHDI. We used regression analysis to find correlation among the different factors and the SHDI values. Out of many factors that

* Corresponding Author's Email: gajwanijuhi@gmail.com.

we analyzed, we concluded that certain factors had a strong correlation. These were educational budget allocated to each state, availability of electricity in households, percentage of population consuming tobacco in each state and health budget allocated to each state. Later on, in the chapter we have discussed how SHDI data can be used to predict other national aggregates such as infant mortality rate, maternal mortality rate and total fertility rate using ensemble machine learning techniques. Our conclusion from the research we conducted is that the subnational human development index is not only a powerful indicator of socio-economic development but is also a predictor of infant mortality rate, maternal mortality rate and total fertility rate for each state within India.

Keywords: subnational human development index, machine learning, education, health, standard of living, ensemble machine learning

INTRODUCTION

Human Development Index (HDI), published by the united nations, is a measure of social and economic development of a nation. It takes into account three dimensions viz., health, education and standard of living that together indicate a nation's socio-economic development.

Now, HDI although provides insights into a country's development but in doing so it aggregates the data of the sub regions within the countries and blurs the regional disparities. So, to overcome these shortcomings of the HDI, another index called the Subnational Human Development Index (SHDI) has been introduced. This index provides data individually about regions within a country and can be used to examine the disparities that exist therein. So, SHDI can be used for the analysis of states within the South Asian country of India. In this chapter, we are primarily be analyzing the SHDI of states within India and the determinants and indicators of the SHDI. The SHDI data being analyzed is from 1990 to 2017. The database consists of information about 161 countries and a total of 1625 regions within those countries.

In order to determine the health component of the SHDI, the key indicator used is the life expectancy of people in a region. Life expectancy is the average number of years a person is expected to live from the time of his/her birth. Similarly, a key indicator of the standard of living is the gross national income per capita. For education, expected years of schooling and mean years of schooling are the two indicators. Expected years of schooling is the number of years of schooling a child is expected to receive. Mean years of schooling is the average number of years of schooling of a person aged 25 years or above.

Many studies have been conducted on the human development index data of several countries across the globe. These studies aim to find correlations between different types of data such as incidence and mortality rates of various types of cancer and the HDI data. Other studies try to analyze the HDI data to find patterns of any chronic diseases, their incidence and mortality.

SHDI metric effectively quantifies the socio-economic development of a nation and can also be used to deduce several other national aggregates. Unlike the HDI, it does not blur the regional disparities within a nation. SHDI can be used to predict a wide range of aggregates such as the incidence and mortality of prostate cancer, the proportion of anemia associated with iron deficiency in different countries, or other metrics such as infant mortality rate, maternal mortality rate. In order to predict these trends using the SHDI data we need an effective machine learning technique. Ensemble machine learning techniques are a group of techniques that increase the prediction accuracy by aggregation. These techniques either utilize the results of more than one prediction model or train multiple instances of a single type of model with different subsets of the data and aggregate the results for the final prediction. So, these techniques can use the dimensions of SHDI as input data to predict any objective function. Prior to prediction, it is important to measure the correlation of the variable to be predicted with the SHDI values. We only predict those variables that have a strong correlation. Also, the other factors used, apart from SHDI values, for prediction, must be highly correlated with the SHDI and its dimensions.

In this chapter, we have conducted research on the possible determinants of the SHDI and the insights or predictions we can make using this data. For health, we considered several factors such as, the number of hospitals in the state, health budget allocated to the state and the percentage of tobacco users in the state. For education, the factors we considered were, budget allocated to each state for education, gross enrolment rate defined as the total number of students who enrolled in school at different grade levels. For standard of living, we considered several factors such as, percentage of households with availability of toilets, electricity and drinking water supply. We plotted a heat map to understand which factors are highly correlated. The regression analysis revealed how the different variables change with respect to each other. We have observed that the educational budget for each state, health budget for each state, percentage of population consuming tobacco in each state and the availability of electricity in each state are powerful determinants of education, health and standard of living respectively in the SHDI index. On the other hand, infant mortality rate, maternal mortality rate and total fertility rate for each state within India can be predicted using the SHDI.

Furthermore, we also tried to understand and derive some factors that might influence the SHDI dimensions, namely, education, health and standard of living in some way. We considered a number of factors targeting different components or dimensions of the SHDI data and boiled down to some factors that were found to have strong correlation. The selected factors include educational budget allocated to each state, availability of electricity in households, percentage of population consuming tobacco in each state and health budget allocated to each state. These findings might aid the policy makers in making more informed decisions.

Related Work

Although the subnational human development index data has not been analyzed much, but there are a few researches on the human development index data. One such work is done by Sanusi [1] in which the author has used the HDI data to measure the degree of deprivation in the urban households in Minna, Nigeria. The work includes examining the housing facilities, housing adequacy, housing space and solid waste disposal as these are some issues that affect human development and are not entirely covered within the HDI. The study states that, while the availability of these amenities might indicate human development, but their absence must indicate some form of depravity. Another study conducted by Lee et al. [2], determines the predictions that can be made using the HDI data. It says that infant and maternal mortality rates of a country can be predicted using the HDI data. A research conducted by Smit Shah [3], highlights some of the determinants of HDI. The author states that GDP per capita, literacy rate, life expectancy at birth, gini index, fertility rate and CO2 emission are significant in empirical regression analysis of the HDI data. Another study conducted by Zangiacomi Martinez & Roza [7], aims to find patterns in the adolescent pregnancy rates and the live births with respect to the human development index data.

Other data such as, the incidence and mortality rates of several medical conditions such as anemia, breast cancer, ovarian cancer, prostate cancer and colorectal cancer have been analyzed with respect to the human development index in several parts of the world. Such studies intend to find patterns between the occurrence of such chronic diseases and the mortality rates in different places of the world due to such medical conditions with respect to the human development index data. One such study conducted by Petry et al. [8] aims to analyze the data of anemia primarily caused by the deficiency of iron in the three categories of countries i.e., low, medium or high human development index. Another research conducted by Fidler et al. [9] established a strong positive correlation between a total of 27 major types of cancer incidence and human development index in 184 countries across the world. A study on incidence and mortality related to prostate cancer and its correlation with the human development index was conducted by Hassanipour-Azgomi et al. [10] in the Asian region in 2012. A positive correlation of 0.475 was seen between standardized incidence rate of prostate cancer and the human development index. Also, a negative correlation of 0.160 was seen between standardized mortality rate of prostate cancer and human development index. Another research on the correlation of colorectal cancer and human development index in Iran was conducted by Enayatrad et al. [11] in 2017 which concluded that the incidence of colorectal cancer was high in regions with a higher human development index.

Research on incidence and mortality rates of breast cancer and their correlation with the human development index in the world was conducted by Ghoncheh, M. et al. [12] in 2012. The study indicated that there is a strong and direct correlation between the standardized incidence rate and human development index. Furthermore, there was a direct correlation between the age-specific mortality rate and the components of Human Development Index. Another research on breast cancer survival was made by Joko-Fru et al. [13] in sub-saharan Africa. A study conducted by Razi et al. [14] aims to find the correlation between the incidence and mortality estimates of ovarian cancer and human development index. The highest standardized incidence rate of ovarian cancer was observed in Singapore, Kazakhstan, and Brunei respectively. the highest standardized mortality rate was observed in the regions of Indonesia, Brunei, and Afghanistan.

A research conducted by Dumuid et al. [15] on the level of physical activity of children and its impact on their health. The study concluded that children residing in countries that have relatively greater HDI values have profound positive relationships between their quality of life and the physical activity they undergo, than those from lower human development index countries. Another study conducted by Takahashi et al. [16] aims to determine whether there is a correlation between the Human Development Index and Inflammatory Bowel Diseases in Brazil. The correlation was found to be moderately positive. A research conducted by Veisani et al. [17] aims to find the correlation between global incidence and mortality rates in pancreatic cancer and the human development index. The study concluded that there are disparities in the pancreatic cancer incidence and mortality rates across the globe. Developed countries have higher rates of pancreatic cancer. Apart from all the components of the HDI mean years of schooling had the greatest effect on the disparities pertaining to the rate of pancreatic cancer.

METHODOLOGY

Data Sets

The subnational human development index is a measure of the socio-economic development of a particular region within a country. The SHDI dataset is available on https://www.nature.com/articles/sdata201938 [18]. The information in the dataset is collected over a period from 1990 to 2017 from 161 countries and 1625 regions within those countries. The SHDI is a composition of three dimensions, namely, health, education and standard of living. Each component has an indicator through which it is determined. The SHDI value is calculated by taking an average of the three dimensions.

The education dimension has two indicators, viz., mean years of schooling of adults of age 25 or more and expected years of schooling for a child. The health dimension has an indicator called life expectancy, defined as average number of years a person is expected to live. The standard of living dimension has an indicator called the gross national income per capita defined as the income made by a person in a region annually. The dimensions are calculated from the indicators as follows:

Dimension Index = (actual value – min value)/(max value – min value)

The SHDI values are calculated from the dimension indices by taking their geometric mean.

Apart from the indicators used for the calculation of the SHDI, this research aimed to find the other indicators that might influence the value of SHDI. Our primary data source is the open government data platform (OGD) available on "https://data.gov.in/." It contains all the data used for analysis from the different states within India. Our approach was to find the indicators of the SHDI as well as the predictions that can be made from the SHDI data. During the study, we did correlation and regression analysis in order to discover the relationship between the SHDI and the independent variables, such as, availability of electricity in each state, the budget for education allocated to each state, internet availability in each state etc.

Another aspect of this study was to discover the factors that can be predicted from the SHDI data. In order to do so, we isolated a number of factors that might be remotely related to any of the components of the SHDI, such as, infant mortality rate, maternal mortality rate, total fertility rate etc. We conducted regression and correlation analysis in order to determine the relationship between them and the SHDI. Then, we used ensemble machine learning techniques in order to predict the values of infant mortality rate, maternal mortality rate and total fertility rate respectively using the SHDI data. Infant Mortality Rate (IMR) is defined as the number of children who die before the age of one year per 1000 live children born. Maternal Mortality Rate (MMR) is defined as the number of women who die per 100,000 birth procedures. Total Fertility Rate (TFR) is measured by the average number of children per woman.

For the first part of our research, we identified a number of factors that could act as indicators for the SHDI. In order to find out whether actually those factors are correlated with the SHDI we plotted graphs between the two. We considered several factors for each of the dimension in the SHDI. For health, we considered several factors such as, the number of hospitals in the state, health budget allocated to the state and the percentage of tobacco users in the state. For education, the factors we considered were, budget allocated to each state for education, gross enrolment rate defined as the total number of students who enrolled in school at different grade levels. For standard of living, we considered several factors such as, percentage of households with availability of toilets, electricity and drinking water supply. Other factors that we considered were unemployment ratio in each state, the forest cover in each state, number of users of wireless internet. Different levels of correlation were observed for different attributes. For this, we plotted a heat map to understand which factors are highly correlated. The regression analysis revealed how the different variables change with respect to each other. Several plots revealed positive and negative linear relationships as shown in Table 1.

Table 1. Relationship between different indicators of SHDI

Independent Variable	Dependent Variable	Relationship
Number of Hospitals	Life Expectancy	Very Low Correlation
Health Budget	Life Expectancy	Very Low Correlation
Expected Years of Schooling	Life Expectancy	High Correlation-Positive Linear Relationship
Mean Years of Schooling of Adults	Life Expectancy	High Correlation-Positive Linear Relationship
Gross National Income Per Capita	Life Expectancy	High Correlation-Positive Linear Relationship
Percentage of Tobacco Users	Life Expectancy	Medium Correlation-Negative Linear Relationship
Gross National Income Per Capita	Education Index	High Correlation-Positive Linear Relationship
Gross Enrolment Rate	Education Index	Very Low Correlation
Educational Budget	Education Index	Very Low Correlation
Toilet	Gross Income Per Capita	Very Low Correlation
Drinking Water Supply	Gross Income Per Capita	Low Correlation
Electricity Supply	Gross Income Per Capita	High Correlation-Positive Linear Relationship
Expected Years of Schooling	Gross Income Per Capita	High Correlation-Positive Linear Relationship
Mean Years of Schooling of Adults	Gross Income Per Capita	High Correlation-Positive Linear Relationship

After analyzing the relationships between these indicators, we can select the factors that have a high correlation with the SHDI and use them to predict the infant mortality rate, maternal mortality rate and total fertility rate.

The consolidated dataset that we used for prediction consisted of 35 rows, consisting of all the states and union territories of India. We chose attributes for prediction by referring the correlation analysis between indicators. The attributes used for prediction include, SHDI, educational budget for each state, percentage of tobacco users in each state and availability of electricity in each state. For health dimension, we have identified a new indicator i.e., percentage of tobacco users in each state. For education dimension, we have identified a new indicator i.e., educational budget for each state. For standard of living we have identified a new indicator i.e., availability of electricity in each state. A screenshot of the dataset we used is shown in Figure 1.

	A	B	C	D	E	F	G	H	I	J	K	L	M	N	O	P	Q	R
1	State	shdi	healthind	incindex	edindex	lifexp	lgnic	esch	msch	num_hosp	health_bu	tobacco_u	GER	edidx	edu_budg	toilet	drinking_v	electricity
2	Andaman	0.702143	0.852964	0.669321	0.607179	75.43929	9.036429	12.97143	7.407143	5	3230154	30.33871	88.93	40.28	39.06	60.2	67.87	98.5
3	Andhra Pr	0.519464	0.691143	0.522357	0.390357	64.925	8.062857	9.535714	3.764286	1380	64593685	20	84.48	67.88	49.27	35.8	55.08	97.5
4	Arunachal	0.546821	0.694321	0.536571	0.443321	65.13214	8.157857	10.6	4.467857	10	8932825	45.5	126.76	24.64	18.3	83.9	70.02	98.5
5	Assam	0.511643	0.668643	0.464929	0.432357	63.46429	7.683929	9.607143	4.964286	183	46168315	48.2	106.11	60.29	62.81	86.4	45.86	94.6
6	Bihar	0.464321	0.655536	0.43875	0.349857	62.61071	7.509286	8.332143	3.55	1039	78436304	25.9	107.67	42.05	35.16	20.1	58.76	79.4
7	Chandigar	0.658464	0.815679	0.631429	0.556929	73.01429	8.786071	12.75357	6.078571	60	3723457	13.7	81.44	82.9	69.54	90.4	97.73	98.5
8	Chattisgar	0.572143	0.730071	0.507321	0.507179	67.46786	7.963929	12.82143	4.532143	445	41087237	39.1	100.02	46.55	57.53	17.6	54.456	96.7
9	Dadra anc	0.683	0.802	0.607286	0.656107	72.12857	8.624643	14.36071	7.714286	18	1220127	30.33871	82.5	58.99	33.54	46.7	45.57	97.4
10	Daman an	0.6715	0.808357	0.659429	0.569107	72.54643	8.970714	11.575	7.428571	5	578840	30.33871	82.03	41.13	51.96	68.1	71.42	100
11	Goa	0.656536	0.777679	0.642821	0.568321	70.54643	8.861071	11.90714	7.128571	83	8695581	9.7	102.57	48.96	66.41	92.6	95.78	98.6
12	Gujarat	0.557964	0.692179	0.577357	0.435464	64.98929	8.428214	9.485714	5.160714	3807	82590504	25.1	97.24	58.37	34.94	63.9	43.41	97.3
13	Haryana	0.576964	0.708857	0.585143	0.464286	66.075	8.479286	10.63929	5.060714	1669	41452045	23.6	91.41	61.95	66.57	32.6	69.78	99
14	Himachal	0.607714	0.742286	0.582821	0.52075	68.25357	8.464286	11.98571	5.635714	485	20267635	16.1	98.8	50.53	62.87	54.7	74.32	98.3
15	Jammu an	0.575786	0.728036	0.566143	0.46525	67.32857	8.353929	10.83571	4.928571	96	30726552	23.7	85.98	55.32	54.56	53.5	77.34	99.4
16	Jharkhand	0.570036	0.755036	0.484821	0.507536	69.075	7.815357	12.84286	4.525	478	30725239	38.9	109.22	41.06	49.75	65.1	54.456	97.5
17	Karnataka	0.549893	0.708893	0.5415	0.434321	66.075	8.190714	9.789286	4.871429	2226	68970261	22.8	102.98	43.91	20.95	15.8	54.456	93.9
18	Kerala	0.647643	0.805393	0.583107	0.580821	72.35	8.465357	12.08571	7.353571	890	57425626	12.7	95.44	69.57	39.35	24.8	71.68	97.9
19	Lakshadw	0.712643	0.819929	0.693571	0.637393	73.29286	9.196429	13.41429	7.942857	1	705560	30.33871	73.8	76.63	79.03	94.7	18.89	97.9

Figure 1. Screenshot of the dataset used for prediction.

In the dataset shown in Figure 1, column lifexp refers to the life expectancy, Ignic refers to gross income per capita, esch refers to expected years of schooling and msch refers to mean years of schooling, IMR is infant mortality rate, MMR is maternal mortality rate, TFR is total fertility rate. In order to predict the different rates, we need to discretize their values. So, we have used the five quantile measures to create three categories of values, viz., low (L), medium (M) and high (H). We split the data into training and test data in the proportion of 70 and 30 percent respectively. In order to use ensemble machine learning techniques, we need to train certain basic models such as decision trees, logistic regression etc. Such models are used for prediction in combination in ensemble methods like voting. Finally, we train and test the voting classifier, gradient boosting classifier, ada boost classifier and bagging classifier.

Data Analysis

We conducted experiments using the ensemble machine learning techniques to make the predictions. Ensemble methods either use several homogeneous weak learners i.e., basic classification algorithms such as decision tree and train each of them in separate ways or they use several heterogeneous weak learners and finally aggregate the results of all the learners using different techniques. The different learners can either be trained in parallel or the results of one learner can be used to improve the predictions of the other learners by training them sequentially. There are several types of algorithms to aggregate the results of the weak learners aimed to reduce variance, decrease bias and improve predictions. These methods are bagging, boosting and stacking. The choice of the algorithm used to aggregate the results of the various weak learners depends on the basic algorithms used for classification. For example, deep decision trees are overfitted and hence have high variance, so bagging can be used for aggregating different trees trained separately.

Sequential ensemble methods are those where the basic predictors are generated sequentially i.e., one after another. Successive predictors use the results of the former predictors in order to improve performance.

Parallel ensemble methods are those where the basic predictors are generated in parallel i.e., simultaneously. In such methods, results of several prediction algorithms are combined to improve prediction.

Bagging

Bagging stands for bootstrap aggregation. This is a type of parallel ensemble method where several samples of data are generated and are used to train different models. For example, different trees can be trained using the different subsets of data.

Boosting

Boosting involves creating several predictive models sequentially and using the results of the previous models to improve predictions of successive models.

Voting

It involves generating a number of models (typically of differing types) and using statistical techniques to make the final prediction. In hard voting, we usually choose the results of prediction by simply conducting a majority vote among all the models. Soft Voting averages the probabilities calculated by the different models to arrive at the final prediction.

Random Forest

In random forest, each tree in the group is worked from an example drawn with substitution (for example a bootstrap test) from the preparation set. What's more, rather than utilizing all the highlights, an arbitrary subset of highlights is chosen, further randomizing the tree. Therefore, the bias of the forest increments marginally, however because of the averaging of less corresponded trees, its change diminishes, bringing about a general better model.

ANALYSIS-RELATED RESULTS

Figure 2 shows the correlation between the different indicators using correlation heat maps. In the map, regions with lower contrast indicate strong correlation among the corresponding variables and the regions with higher contrast indicate weak correlation among the corresponding

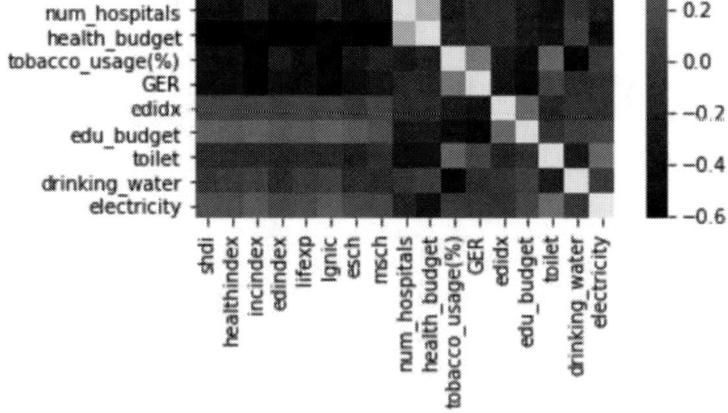

Figure 2. Correlation heat map between the various indicators.

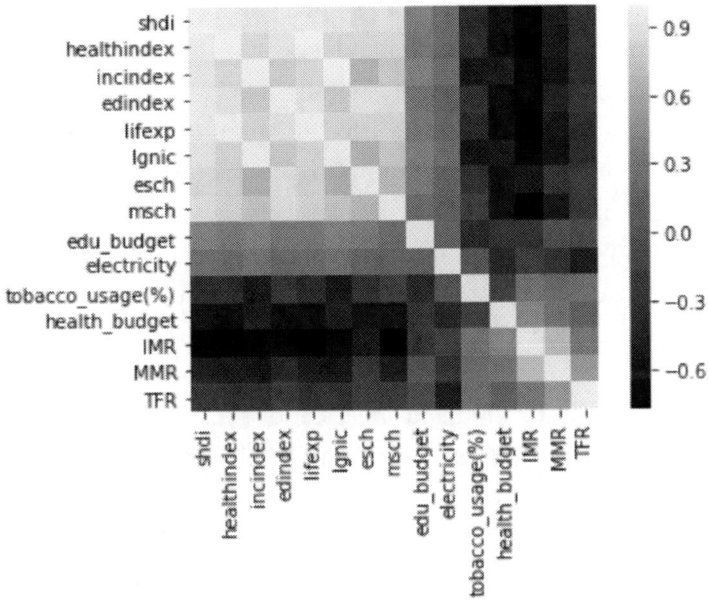

Figure 3. Correlation heat map between the selected indicators and predictors.

variables. For example, a lower contrast between life expectancy and health index indicates strong correlation. Similarly, another heat map in Figure 3 shows correlation among the selected indicators and predictors. For example, a lower contrast between health budget and maternal mortality rate indicates strong correlation.

After identifying the correlations between indicators, we plotted certain graphs between each of the dimensions i.e., health, education and standard of living of SHDI and their indicators to study each correlation closely.

Health Dimension

In order to observe how the number of hospitals in a state impact the average life expectancy of individuals in the state we plotted a graph as shown in Figure 4. We can clearly observe that there is no definite trend between the two indicators. Hence, we could not use the indicator number of hospitals for prediction. Another indicator for life expectancy can be the health budget allocated to each state. So, we plot a graph between the two as shown in Figure 5 to analyze their relationship. In this case also we can observe that there is no noticeable trend between the two and hence the indicator health budget cannot be used for prediction. Another indicator called the expected years of schooling that forms the education dimension can affect the health of people. A plot between the expected years of schooling and the average life expectancy of individuals in each state is shown in Figure 6. There is a linear trend among the two, as the expected years of schooling increase so does the average life expectancy. This means that expected years of schooling would be a good measure for prediction. We can observe similar trend in the case of mean years of schooling in Figure 7. This means that mean years of schooling would also be a good measure for prediction. From the correlation heat map, we can observe gross income per capita is strongly correlated with the average life expectancy of people in each state. As we can see from Figure 8 that there is a linear relationship between gross income per capita and life expectancy.

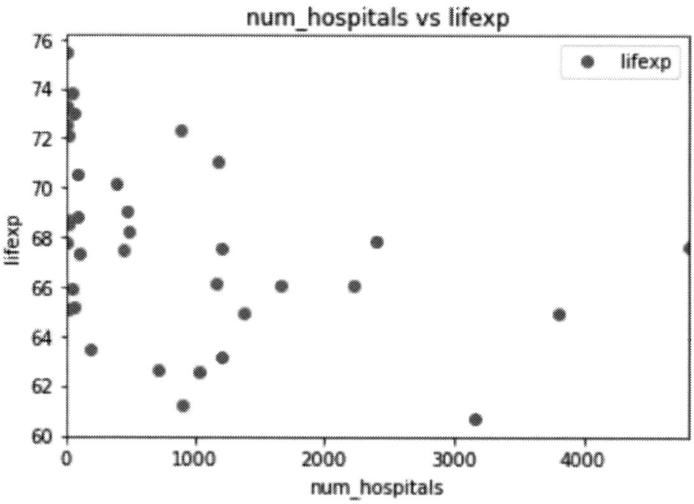

Figure 4. Plot between the number of hospitals in a state and the average life expectancy of individuals in each state.

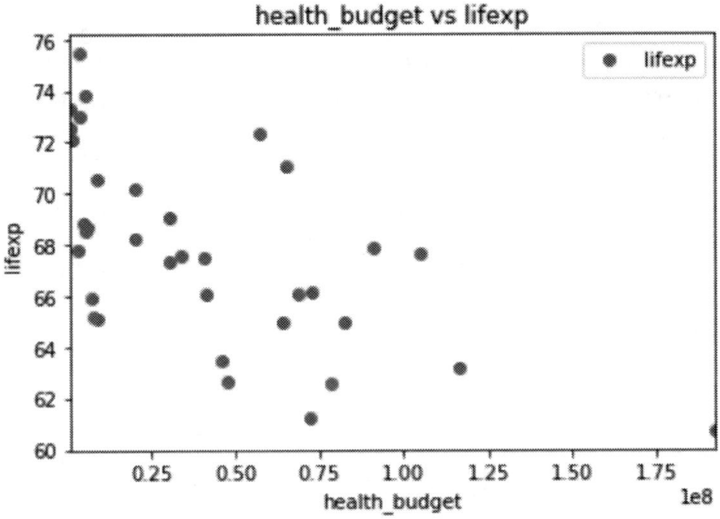

Figure 5. Plot between the health budget allocated to each state and the average life expectancy of individuals in each state.

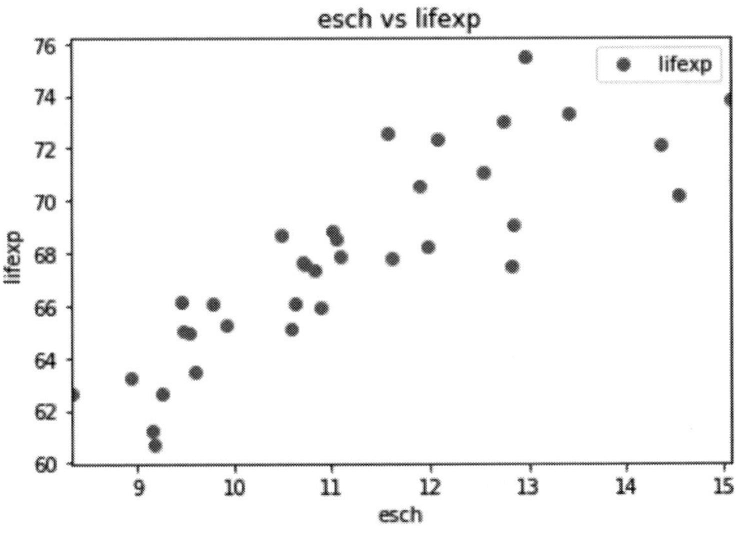

Figure 6. Plot between the expected years of schooling for each state and the average life expectancy of individuals in each state.

Another factor that could affect health, probably negatively, was percentage tobacco consumption in each state. A plot between the two is shown in Figure 9. We can observe a negative correlation between the two, as the percentage of tobacco usage increase, average life expectancy decreases.

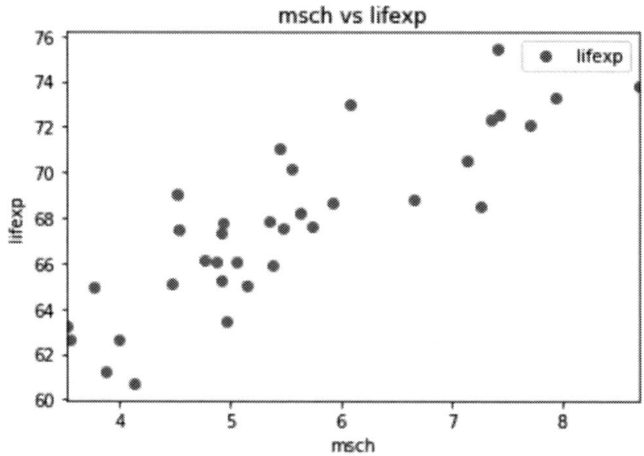

Figure 7. Plot between the mean years of schooling for each state and the average life expectancy of individuals in each state.

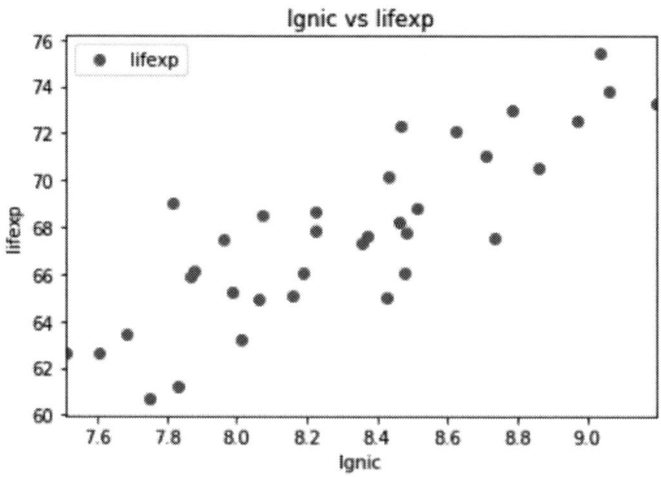

Figure 8. Plot between the average gross national income per capita for an individual in each state and the average life expectancy of individuals in each state.

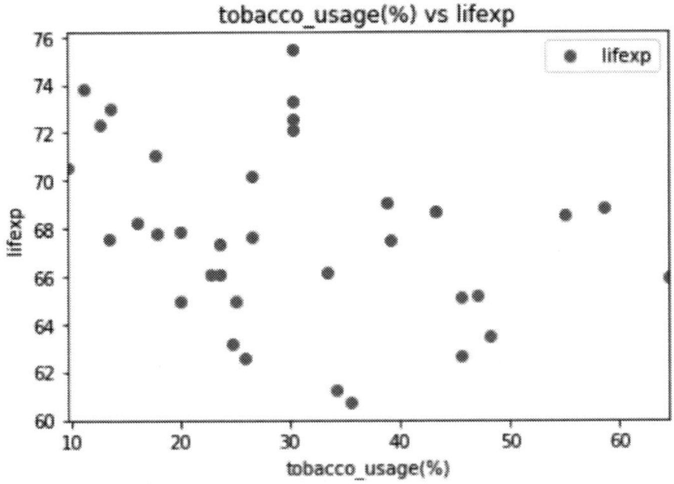

Figure 9. Plot between the average percentage of tobacco usage for an individual in each state and the average life expectancy of individuals in each state.

Education Dimension

For the education dimension we plotted graphs of indicators against the education index. In Figure 10, we have plotted a graph between the average gross income per capita of an individual in a state and the education index of each state. It shows that as the average gross income per capita increases so does the education index.

Gross Enrolment Rate (GER) is defined as the number of students who enroll in different standards in all the schools in each state. In Figure 11, we have a plot between GER and the education index of each state. We observe that there does not seem to be a direct correlation between the two.

We plotted another graph between the education budget allocated to each state and the education index in Figure 12. We came to the conclusion that higher the budget allocated, higher is the education index for each state. But at the same time, there are several outliers that do not justify our conclusion.

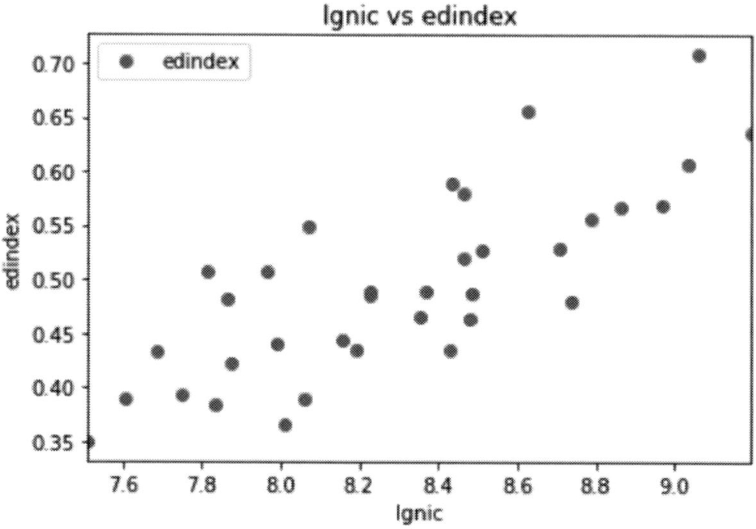

Figure 10. Plot between the average gross national income per capita for an individual in each state and the education index of each state.

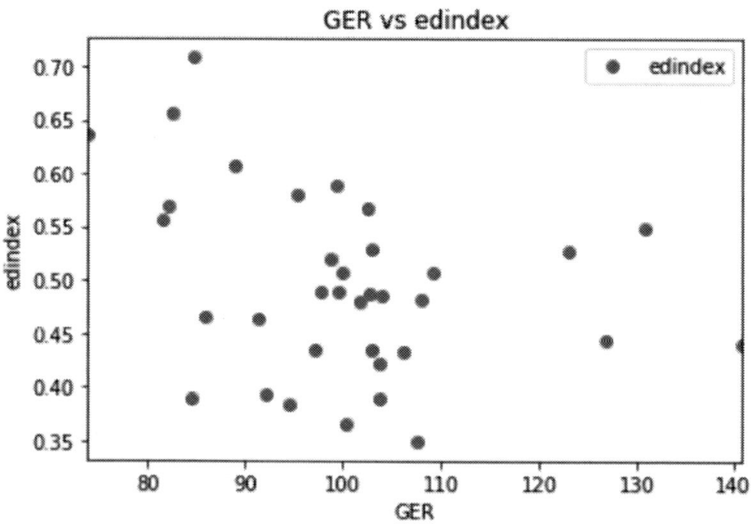

Figure 11. Plot between GER of each state and the education index of each state.

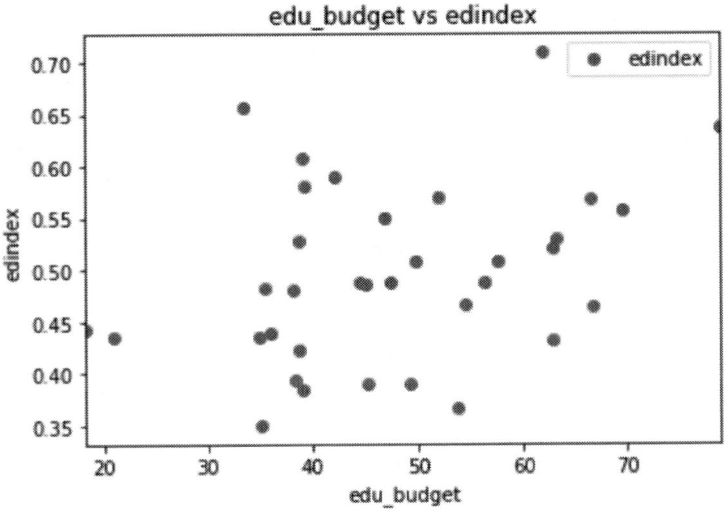

Figure 12. Plot between education budget allocated to each state and the education index of each state.

Standard of Living Dimension

Standard of living is measured using average gross national income per capita for an individual in each state. An index called incindex is calculated using the formula discussed in methodology section to calculate the dimension index. In order to identify more indicators, we plotted graphs of different factors against the incindex. First, we plotted a graph of availability of toilets within dwelling units in each state against the incindex. It can be observed from Figure 13 that there is no definite pattern or correlation between the two.

Another factor that can be related to standard of living is availability of drinking water in households in each state. In Figure 14., we observe a plot between the availability of drinking water in households in each state against the incindex, we cannot draw a direct relationship between the two.

Availability of electricity in households indicates standard of living. In the Figure 15, we can observe that although there are certain outliers, but as the availability of electricity in households increases so does the standard of living.

The level of education in the population can indicate the standard of living, so we plot a graph between the two in Figure 16. It can be observed from the plot that as the edindex increases so does the incindex.

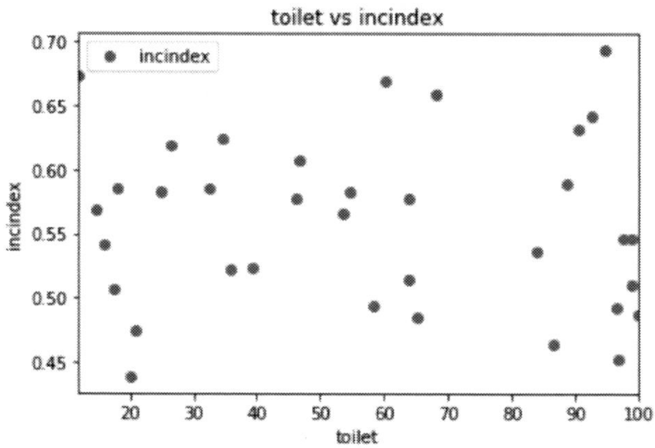

Figure 13. Plot between the availability of toilets within dwelling units in each state against the incindex.

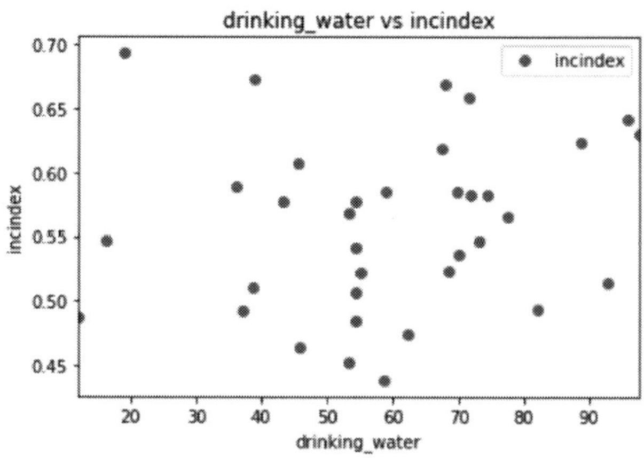

Figure 14. Plot between the availability of drinking water in households in each state against the incindex.

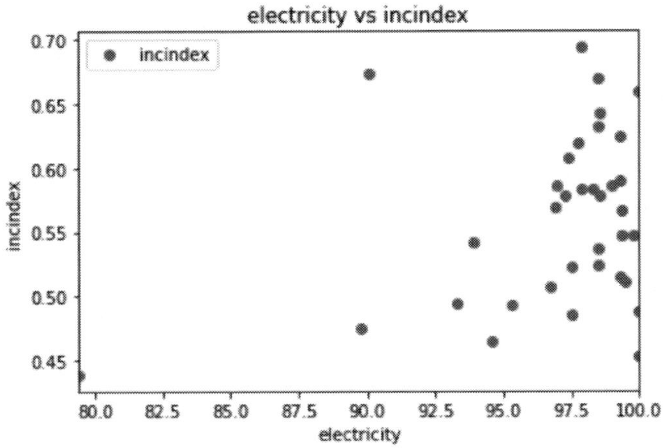

Figure 15. Plot between availability of electricity in households and incindex.

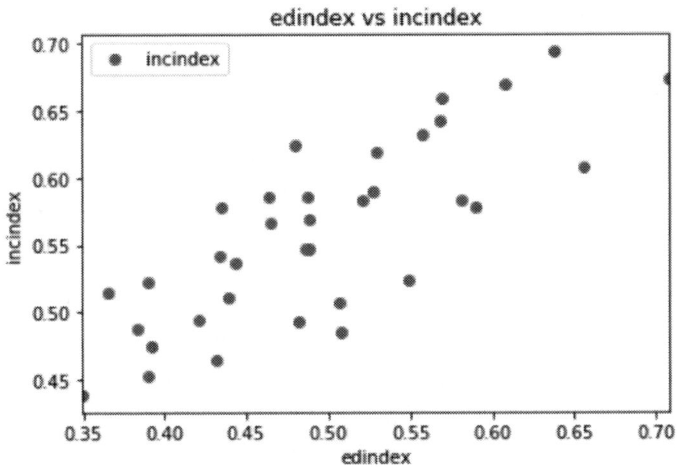

Figure 16. Plot between education index and incindex.

PREDICTION-RELATED RESULTS

After identifying the different indicators, we tried to predict IMR, MMR and TFR.

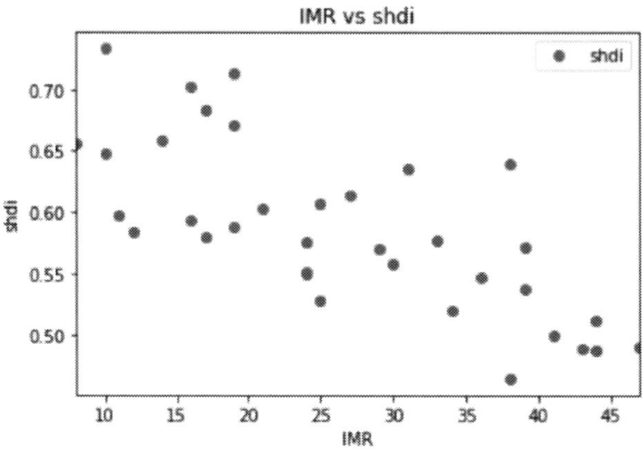

Figure 17. Plot between IMR and SHDI.

IMR Predictions

Infant Mortality Rate (IMR) is defined as the number of children who die before the age of one year per 1000 live children born. In order to predict the IMR, we first plotted a graph between IMR and SHDI to observe the trend between the two as shown in Figure 17. We can clearly observe that as the SHDI values decrease so does the IMR. Since the IMR is varying in accordance with the SHDI values, so we can predict its value for all the states using the SHDI data. For prediction, we used the ensemble machine learning techniques. We applied all the ensemble techniques one by one to determine which method gave the maximum accuracy of prediction. We used the following categories of data or attributes in order to predict the IMR values: subnational human development index values, educational budget allocated to each state, availability of electricity in households, percentage of population consuming tobacco in each state and health budget allocated to each state. In order to train the models, we discretized the IMR values into three categories, namely, low (L), medium (M) and high (H). In python, we used the scikit learn library. First, we created a numpy array from a data frame with all the attributes used for prediction named X. Then, the IMR values were placed in another numpy array named Y. We then used the train_test_split function in order to split the data into training and testing subsets. Here, we divided the data in the ratio of 70:30 for training and testing respectively. In the dataset there are a total of 35 states, so data from about 11 states would be used for prediction. Now, in order to use any ensemble technique such as bagging, boosting or stacking we first need to train basic weak learners such as decision trees etc. So, we used DecisionTreeClassifier class from sklearn.tree with gini index as the attribute selection criteria. The number of misclassified samples from this technique were 3. Next, we experimented with DecisionTreeClassifier with entropy as the attribute selection criteria. Using this technique, the number of misclassified samples jumped to 4. Another basic model used was LogisticRegression which led to a total of 4 misclassified samples. An Ensemble model called RandomForestClassifier was used which led to a total of 5 misclassified samples. Next, we used another ensemble technique called voting in which we use several heterogeneous basic classifiers, and then combine their results using two different techniques called soft and hard voting respectively. For voting, we used three classifiers, namely, LogisticRegression, RandomForestClassifier and DecisionTreeClassifier with entropy as the attribute selection criteria. Finally, we used VotingClassifier to combine their results. The final result was 3

Table 2. IMR prediction results

Classifier	Number of Misclassified Samples	Accuracy (%)
Decision Tree-Gini Index	3	73
Decision Tree-Entropy Index	4	64
Logistic Regression	4	64
Boosting-Gradient	4	64
Boosting-Ada	3	73
Voting-Hard	3	73
Voting-Soft	3	73
Voting-Weighted	3	73

misclassifications. Next, we experimented with the AdaBoostClassifier and GradientBoostingClassifier leading to a total of 3 and 4 misclassifications respectively. Lastly, we experimented with the BaggingClassifier with a total of 5 misclassifications. So, we can say we can predict the IMR with an accuracy of about 73%. The accuracy can be improved if we consider a larger dataset such as data from more than a single country.

From Table 2, we can clearly see that the ensemble methods provide better results than the basic learners.

MMR Predictions

Maternal Mortality Rate (MMR) is defined as the number of women who die per 100,000 birth procedures. MMR clearly varies with respect to the SHDI values as shown in Figure 18. In order to predict the MMR values based on classification, we first need to discretize the original values, i.e., convert the continuous data into categorical data. We achieved this by using the pandas library's categorical data type. Pandas.cut divides by using user defined bins and pandas.qcut generates quantile based discretization. This technique led to division of MMR values into four categories, namely, very low, low, high and very high. We used the following attributes for prediction: subnational human development index values, educational budget allocated to each state, availability of electricity in households, percentage of population consuming tobacco in each state and health budget allocated to each state. For prediction, we used the ensemble machine learning techniques. We applied all the ensemble techniques one by one to determine which method gave the maximum accuracy of prediction.

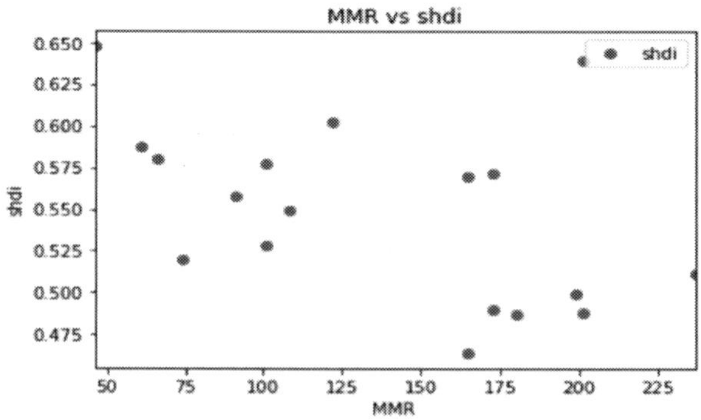

Figure 18. Plot between MMR and SHDI.

In python, we used the scikit learn library. First, we created a numpy array from a data frame with all the attributes used for prediction named X. Then, the IMR values were placed in another numpy array named Y. We then used the train_test_split function in order to split the data into training and testing subsets. Here, we divided the data in the ratio of 70:30 for training and testing respectively. In the dataset there are a total of 35 states, so data from about 11 states would be used for prediction. Now, in order to use any ensemble technique such as bagging, boosting or stacking we first need to train basic weak learners such as decision trees etc. So, we used DecisionTreeClassifier class from sklearn.tree with gini index as the attribute selection criteria. The number of misclassified samples from this technique were 3. Next, we experimented with DecisionTreeClassifier with entropy as the attribute selection criteria. Using this technique, the number of misclassified samples jumped to 4. Another basic model used was LogisticRegression which led to a total of 2 misclassified samples. Next, we experimented with the class called GaussianNB which represents Gaussian Naïve Bayes classifier resulted in a total of 4 misclassifications. An Ensemble model called RandomForestClassifier was used which led to a total of 4 misclassified samples. Next, we used another ensemble technique called voting in which we use several heterogeneous basic classifiers, and then combine their results using two different techniques called soft and hard voting respectively. For voting, we used three classifiers, namely, LogisticRegression, RandomForestClassifier and DecisionTreeClassifier with entropy as the attribute selection criteria. Finally, we used VotingClassifier to combine their results. The final result was 3 misclassifications. Next, we experimented with the AdaBoostClassifier and GradientBoostingClassifier leading to a total of 4 misclassifications in both the cases. Lastly, we experimented with the BaggingClassifier with a total of 5 misclassifications. So, we can say we can predict the MMR with an accuracy of about 82%. The accuracy can be improved if we consider a larger dataset such as data from more than a single country. From Table 3, we can clearly see that the ensemble methods provide better results than the basic learners.

Table 3. MMR prediction results

Classifier	Number of Misclassified Samples	Accuracy (%)
Decision Tree-Gini Index	3	73
Decision Tree-Entropy Index	4	64
Logistic Regression	2	82
GaussianNB	4	64
Boosting-Gradient	4	64
Boosting-Ada	4	64
Voting-Hard	3	73
Voting-Soft	3	73
Voting-Weighted	3	73
Random Forest	4	64

TFR Predictions

Total Fertility Rate (TFR) is measured by the average number of children per woman. TFR clearly varies with respect to the SHDI values as shown in figure 19. In order to predict the MMR values based on classification, we first need to discretize the original values, i.e., convert the continuous data into categorical data. We achieved this by using the pandas library's categorical data type. Pandas.cut divides by using user defined bins and pandas.qcut generates quantile based discretization. This technique led to division of TFR values into four categories, namely, very low, low, high and very high. We used the following attributes for prediction: subnational human development index values, educational budget allocated to each state, availability of electricity in households, percentage of population consuming tobacco in each state and health budget allocated to each state. For prediction, we used the ensemble machine learning techniques. We applied all the ensemble techniques one by one to determine which method gave the maximum accuracy of prediction.

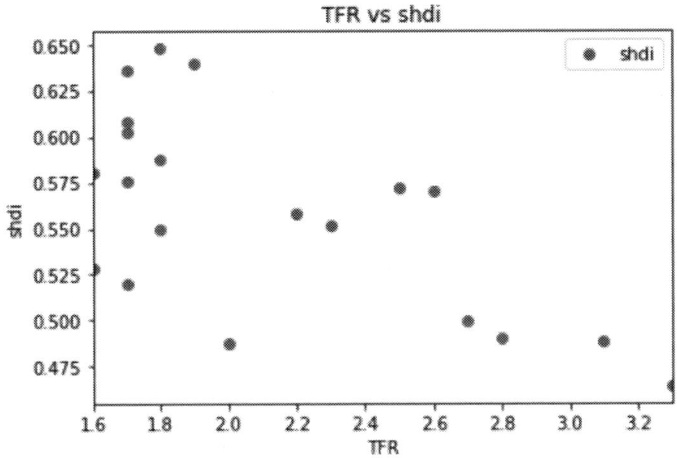

Figure 19. Plot between TFR and SHDI.

In python, we used the scikit learn library. First, we created a numpy array from a data frame with all the attributes used for prediction named X. Then, the IMR values were placed in another numpy array named Y. We then used the train_test_split function in order to split the data into training and testing subsets. Here, we divided the data in the ratio of 70:30 for training and testing respectively. In the dataset there are a total of 35 states, so data from about 11 states would be used for prediction. Now, in order to use any ensemble technique such as bagging, boosting or stacking we first need to train basic weak learners such as decision trees etc. So, we used DecisionTreeClassifier class from sklearn.tree with gini index as the attribute selection criteria. The number of misclassified samples from this technique were 4. Next, we experimented with DecisionTreeClassifier with entropy as the attribute selection criteria. Using this technique, the number of misclassified samples came down to 3. Another basic model used was LogisticRegression which led to a total of 3 misclassified samples. Next, we experimented with the class called GaussianNB which represents Gaussian Naïve Bayes classifier resulted in a total of 3 misclassifications. An Ensemble model called RandomForestClassifier was used which led to a total of 3 misclassified samples. Next, we used another ensemble technique called voting in which we use several heterogeneous basic classifiers, and then combine their results using two different techniques called soft and hard voting respectively. For voting, we used three classifiers, namely, LogisticRegression, RandomForestClassifier and DecisionTreeClassifier with entropy as the attribute selection criteria. Finally, we used VotingClassifier to combine their results. The final result was 3 misclassifications. Next, we experimented with the AdaBoostClassifier and GradientBoostingClassifier leading to a total of 4 and 5 misclassifications respectively. Lastly, we experimented with the BaggingClassifier with a total of 4 misclassifications. So, we can say we can predict the TFR with an accuracy of about 73%. The accuracy can be improved if we consider a larger dataset such as data from more than a single country. The results are summarized in Table 4.

Table 4. TFR Prediction Results

Classifier	Number of Misclassified Samples	Accuracy (%)
Decision Tree-Gini Index	4	64
Decision Tree-Entropy Index	3	73
Logistic Regression	3	73
GaussianNB	3	73
Boosting-Ada	4	64
Voting-Hard	3	73
Voting-Soft	3	73
Voting-Weighted	3	73
Random Forest	3	73

DISCUSSION

There can be many implications of this work. It can be used to make certain estimates directly. For example, with only the knowledge of SHDI values, one can estimate IMR. It can provide a whole new perspective to the policy makers and make their decisions fact-driven. For instance, with the knowledge of IMR estimates for each district, one can take appropriate measures to scale the health facilities accordingly. More funds can be allocated to states with higher mortality rates that can lead to fast paced development of health facilities in such districts.

Another implication of this research could be the factors that influence SHDI dimensions and in turn affect the overall socio-economic development. By identifying such factors, we can explore new avenues towards the overall development of a region. For instance, lets consider the impact of availability of electricity on the standard of living. This work shows high correlation between the two and thus making it an important factor towards socio-economic development.

There are certain limitations to this study. We can improvise in some ways to alleviate them. With the aim of improving the accuracy, one can experiment with a wider dataset such as considering more than one country's SHDI data. Other than that, one could try a different set of algorithms for prediction which might cause the accuracy to change. Experimenting with different factors for each dimension might lead to an entirely different set of results.

CONCLUSION

In this chapter, we analyzed subnational human development index data and its indicators as well as dimensions. We tried to find the relationship between the subnational human development index data and other national level aggregates such as infant mortality rate (IMR), maternal mortality rate (MMR) and total fertility rate (TFR). We also tried to understand and derive some factors that might influence the SHDI dimensions, namely, education, health and standard of living in some way. For health, we considered several factors such as, the number of hospitals in the state, health budget allocated to the state and the percentage of tobacco users in the state. For education, the factors we considered were, budget allocated to each state for education, gross enrolment rate defined as the total number of students who enrolled in school at different grade levels. For standard of living, we considered several factors such as, percentage of households with availability of toilets, electricity and drinking water supply. We plotted a heat map to understand which factors are highly correlated. The regression analysis revealed how the different variables change with respect to each other. Finally, we came to the following conclusions: (a) the educational budget allocated to each state positively affects the education index in the SHDI data. (b) the health budget allocated to each state positively affects the life expectancy in the SHDI data. (c) the tobacco consumption in each state negatively impacts the life expectancy in the SHDI data. (d) the average availability of electricity in households in each state positively impacts the standard of living. (e) conversely, we did not observe any trend between the number of hospitals in a state and the average life expectancy of the people in the state. (f) also, we could not find any particular correlation between the Gross Enrolment Rate (GER) and the education index. (g) availability of toilets, unemployment rate and availability of drinking water within the dwelling units also could not be correlated with the standard of living dimension directly. (h) IMR varies negatively with the SHDI values and hence can be predicted using the SHDI data with an accuracy of 73% (i) MMR varies negatively with the SHDI values and hence can be predicted using the SHDI data with an accuracy of 82% (j) TFR varies negatively with the SHDI values and hence can be predicted using the SHDI data with an accuracy of 73%.

REFERENCES

[1] Yekeen A. Sanusi: "Application of human development index to measurement of deprivations among urban households in Minna, Nigeria," in *Habitat International,* volume 32, issue 3, September 2008, pages 384-398.

[2] Kwang-sun Lee, Sang-chul Park, Babak Khoshnood, Hui-Lang Hsieh and Robert Mittendorf: "Human development index as a predictor of infant and maternal mortality rates," in *The Journal of Pediatrics,* volume 131, issue 3, September 1997, pages 430-433.

[3] Shah, Smit: "Determinants of Human Development Index: A Cross-Country Empirical Analysis."

[4] Juhi Gajwani & Pinaki Chakraborty: "Students' Performance Prediction using Feature Selection and Supervised Machine Learning Algorithms," in 3rd International Conf. Innovative Computing and Communication (ICICC-2020), page 31.

[5] Sudhir Anand & Amartya K. Sen: "Human Development Index: Methodology and Measurement," in New York: Human Development Report Office Occasional Paper 12; 1994.

[6] Ambuj D Sagar a & Adil Najam b': "The human development index: a critical review," in *Ecological Economics* 25(3):249-264 · June 1998.

[7] Edson Zangiacomi Martinez & Daiane Leite da Roza: "Ecological analysis of adolescent birth rates in Brazil: Association with Human Development Index," in *Women and Birth,* volume 33, issue 2, March 2020, pages e191-e198.

[8] Nicolai Petry, Ibironke Olofin, Richard F. Hurrell, Erick Boy, James P. Wirth, Mourad Moursi, Moira Donahue Angel and Fabian Rohner: "The Proportion of Anemia Associated with Iron Deficiency in Low, Medium, and High Human Development Index Countries: A Systematic Analysis of National Surveys," in *Nutrients* 2016, 8(11), 693;

[9] Miranda M. Fidler, Isabelle Soerjomataram and Freddie Bray: "A global view on cancer incidence and national levels of the human development index," in *International Journal of Cancer.*

[10] Hassanipour-Azgomi, S., Abdollah Mohammadian-Hafshejani, Mahshid Ghoncheh, Farhad Towhidi, Saeid Jamehshorani, Hamid Salehiniya: "Incidence and mortality of prostate cancer and their relationship with the Human Development Index worldwide," in *Prostate International,* volume 4, issue 3, September 2016, pages 118-124.

[11] Enayatrad, M., K. Etemad, S. Khodakarim and P. Yavari: "Investigation of the Relationship between the Incidence of Colorectal Cancer and the Human Development Index in Iran: An Ecologic Study," in *Iranian Journal of Epidemiology,* Volume 13, Issue 3 (Vol.13, No.3, Atumn 2017).

[12] Ghoncheh, M. and Mirzaei, M. and Salehiniya, H. (2016): "Incidence and mortality of breast cancer and their relationship with the human development index (HDI) in the world in 2012," in *Asian Pacific Journal of Cancer Prevention,* 16 (18). pp. 8439-8443.

[13] Walburga Y. Joko-Fru, Adalberto Miranda-Filho, Isabelle Soerjomataram, Marcel Egue, Marie-Therese Akele-Akpo, Guy N'da, Mathewos Assefa, Nathan Buziba, Anne Korir, Bakarou Kamate, Cheick Traore, Shyam Manraj, Cesaltina Lorenzoni, Carla Carrilho, Rolf Hansen, Anne Finesse, Ntuthu Somdyala, Henry Wabinga, Tatenda Chingonzoh, Margaret Borok, Eric Chokunonga, Biying Liu, Eva Kantelhardt, Paul McGale, Donald M. Parkin: "Breast cancer survival in sub-Saharan Africa by age, stage at diagnosis and human development index: A population-based registry study," in *International Journal of Cancer.*

[14] Saeid Razi, Mahshid Ghoncheh, Abdollah Mohammadian-Hafshejani, Hojjat Aziznejhad, Mahdi Mohammadian, and Hamid Salehiniya: "The incidence and mortality of ovarian cancer and their relationship with the Human Development Index in Asia," in *ecancer* 10 628.

[15] Dumuid, Dorothea, Carol Maher, Lucy K. Lewis, Tyman E. Stanford, Josep Antoni Martín Fernández, Julie Ratcliffe, Peter T. Katzmarzyk, Tiago V. Barreira, Jean-Philippe Chaput, Mikael Fogelholm, Gang Hu, José Maia, Olga L. Sarmiento, Martyn Standage, Mark S. Tremblay, Catrine Tudor-Locke & Timothy Olds: "Human development index, children's health-related quality of life and movement behaviours: a compositional data analysis," in *Qual Life Res* 27, 1473–1482 (2018).

[16] Takahashi, L. A. R., D. T. Cardial, I. L. Argani, L. R. R. Arnoni, V. N. Cury, L. F. A. C. C. Silva, A. O. Júnior, S. D. F. Borato: "Human Development Index and Inflammatory Bowel Diseases," in *Journal of Advances in Medicine and Medical Research,* 2018, volume 26, issue 2, page 1-8.

[17] Veisani, Y., E. Jenabi, S. Khazaei & Sh. Nematollahi: "Global incidence and mortality rates in pancreatic cancer and the association with the Human Development Index: decomposition approach," in *Public Health,* volume 156, March 2018, pages 87-91.

[18] Smits, Jeroen & Iñaki Permanyer: "The Subnational Human Development Database," in *Sci Data* 6, 190038 (2019).

In: An Introduction to Approaches
Editor: Yi-Tung Chan

ISBN: 978-1-53618-680-2

Chapter 6

QUANTITATIVE TEXTURAL MEASURES OF THE AEROMAGNETIC FIELD: TWO EXAMPLES AT REGIONAL SCALE

Mark Gettings, PhD*
U.S. Geological Survey, Tucson, AZ, US

Abstract

Maps of gravity and magnetic field anomalies (potential fields) often show areas of the map that are coherent in their appearance or texture. These textural patterns in many cases correlate with geologic terranes with distinctive rock types and can be used to extend the geologic terrane beneath cover. In this study the concept of texture of a potential field is defined as the spatial arrangement of the amplitudes in a potential field dataset. Eight quantitative measures of texture are presented as distributions of the measure from a window moving over the data, either in profiles or gridded form. The measures are the frequency and possibility distributions of anomaly amplitude, number of extrema, elongation ratio, mean strike direction, variability of the strike directions, the anomaly surface area, the lacunarity, and the multifractal spectrum. The measures were applied to an aeromagnetic map of the Arctic to quantitatively characterize two different areas of the Arctic Ocean aeromagnetic field and to an area in southeast Arizona to predict the bedrock lithology beneath two areas of alluvial cover.

*Corresponding Author's Email: mgetting@usgs.gov.

PACS: 93.00, 07.05.Kf, 02.70.-c

Keywords: textures of potential fields, gravity, magnetic, textures

AMS Subject Classification: 86-08, 86A20, 86A32

1. Introduction

Maps of gravity and magnetic field anomalies frequently display coherent areas of multiple anomalies (from tens to thousands of anomalies) that show a uniform appearance or texture. When compared with maps of bedrock geology, these areas often correlate directly with lithologic terranes, that is, areas of rock formations characteristic of a particular geologic history or episode. Some examples of lithologic terranes include volcanic terranes and associated rock types formed from a period of volcanism; granitic terranes characterized by large areas of granitic intrusive rocks and the host rocks they intruded; sedimentary terranes of characteristic sequences; and metamorphic terranes composed of metamorphosed versions of older terranes (metavolcanic rocks, metasedimentary rocks, etc.). Lithologic terranes can vary in size from a few to thousands of square kilometers containing many rock types. Sometimes the boundaries of these areas on the geophysical maps are easy to delineate, but often they are gradual and difficult to locate. This is particularly true when attempting to extend geologic mapping from areas of exposed bedrock beneath cover, such as alluvial valleys or plains against a mountain range. Estimating bedrock lithology beneath cover is particularly important in mineral resource exploration because most deposits exposed at the surface have been found and mapped, and discovery of new resources depends on delineating favorable lithologic terranes beneath cover including water and ice as in the Arctic. Thus the working hypothesis is that geologic terranes can be characterized by the distribution and relative shapes (textures) of their aeromagnetic anomalies and these textures can be used to identify geologic terranes beneath cover. This approach is distinct from traditional techniques of interpreting aeromagnetic data by analyzing individual anomalies for their depth and horizontal extents using horizontal gradients, analytic signal, Euler deconvolution and other techniques. This study reports the results of defining quantitative textural measures of aeromagnetic survey data that are geologically meaningful in terms of distribution and shapes of the causative sources in various lithologic terranes.

The term texture derives from textile weaving (Oxford English Dictionary,

1971) and refers to either the surface appearance of or the tactile sensation from touching a textile. For the work described here, texture will refer to the surface appearance. With the advent of digital cameras and images from airborne and spaceborne platforms, digital image processing and pattern recognition became areas of research and development (Weszka et al. (1976), Tuceyran and Jain (1998)). Techniques of image texture delineation were reviewed by Haralick (1979), who defined an image (gray scale) as composed of the two elements tone (degree of grayness) and texture (the spatial variation of tone). With the advent of color images, Shapiro and Stockman (2001) revised this to color and texture, with texture defined as the spatial arrangement of color. Rao (1990) created a taxonomy for the classification of images based on texture. The early applications of image analysis to geoscience topics such as topography, soils, and vegetation include Weszka et al. (1976), Rao (1990) and Renno et al. (1998).

Early applications of textural analysis to potential field data generally used measures from the image processing field applied to aeromagnetic or gravity anomaly grids (Dentith (1995)). Cooper (2004), Cooper and Cowan (2005) and Cooper (2009) continued this approach using traditional image processing textural calculations, with varied success. Gettings (2002) suggested the use of number of extrema and Euclidean length of the data within a moving window along aeromagnetic flight line profile data as measures sensitive to the lithology sensed by the magnetometer. Shankar (2009) compared the Gettings (2002) measures with several from image processing and found them to be the most successful at prediction of the lithology from the aeromagnetic field. Gettings (2009) and Gettings (2012a) applied the measures to gridded data and added four more measures described below. Bultman has used these techniques to extend geologic mapping and mineral exploration beneath cover in southeastern Arizona (Bultman, 2013). Gettings has used these techniques to compare the rock terranes beneath the Colorado Plateau of northern Arizona and southern Utah with that beneath the volcanic cover of southern Nevada (Gettings, 2018). Gettings (2016b) and Gettings (2016a) has presented some of the results of this report at conferences.

2. Textures of Potential Fields

For potential field data (normally magnetic or gravity anomaly field surveys) the field anomaly amplitude is the analog of color and the texture is the spatial arrangement of the amplitudes. The amplitude of a magnetic or gravity anomaly is similar to color in that it is the combination of multiple components. However, the potential field is not subject to singularities as a color image is because of the superposition of the fields of the sources and the fact that the force field decreases as an inverse power of distance. Thus in the case of potential fields one cannot get an infinite slope in amplitude change between two areas as one can in a color image. This is true even though the distribution of source rocks is fractal or multifractal at multiple scales and thus subject to singularities on the boundaries of different lithologic types (Raines, 2002).

Definition: The texture of a potential field is here defined as a description of the spatial variation of amplitudes in a potential field.

3. Textural Measures of Potential Fields

In order to describe the texture, one must find ways of quantifying the spatial variation, that is measures of the texture are needed to quantify the variation. A structural and a statistical approach have been described in the image processing literature (Shapiro and Stockman, 2001). In the structural approach the image is decomposed into primitives called texels (Shapiro and Stockman, 2001) that are regions of the image with consistent tonal properties from which the image is supposed to be composed (Haralick, 1979). The statistical approach uses quantitative measures (for example, a frequency table or histogram) of the arrangement of amplitudes. The statistical approach is used in this work because of the fractal or multifractal distribution of sources generating the potential field. Although the determination of the scaling exponents (dimensions) is straightforward, finding the parameters of a model producing such a distribution is difficult if not impossible (Feder (1988); Gettings (2012b)). Finding such a model is analogous to the structural approach whereas the determination of the multifractal spectrum is analogous to the statistical approach.

In mathematics, a measure on a set is a complete procedure to assign a number to all subsets of that set (see, for example the summary in Wikipedia (2012)). Let X be a set with a nonempty collection Σ of subsets of X including X itself (mathematically, a σ-algebra over X). Then a function μ is a measure if it satisfies the following three properties: non-negativity (1); null empty set

(2); and countable additivity (3).

$$\mu(E) \geq 0 \quad \forall E \in \Sigma \tag{1}$$

$$\mu(\emptyset) = 0 \tag{2}$$

$$\mu(\bigcup_{i \in I} E_i) = \sum_{i \in I} \mu(E_i) \tag{3}$$

where $E_{i \in I}$ is the set of pairwise disjoint sets in Σ. If negative values are allowed, the measure is called a signed measure. Use of unsigned values is preferable as it simplifies calculations. Examples of measures include probability, possibility, various measures of length, area, volume and mass or other physical property distributions. In general, the observed distribution of a measure is used as a robust and data-driven description.

Definition: A textural measure is here defined as the distribution of some measure of the spatial variation in amplitudes of a potential field.

Here measures derived from aeromagnetic data over geologic lithostratigraphic units were used. Equivalent measures can be calculated from gravity anomaly data if available at a sufficiently small and approximately uniform station spacing to provide adequate resolution. The grid interval (spacing between nodes with values in the data grid) determines a low-pass filter on the data because no anomaly can be defined with a characteristic length less than twice the grid interval, and thus some information is always lost in the creation of a grid of the data. In modern aeromagnetic surveys flown with low terrain clearance (typically 50-250 m) and close profile line spacing (typically 100-250 m) the grid can be constructed on a small enough spacing to define most anomalies of interest in the estimation of geologic units. For regional-scale studies larger grid spacing can be used when only larger areas are of interest.

Several textural measures can be computed from the aeromagnetic field. These include the distribution of anomaly amplitude, the number of extrema per unit area, the elongation ratio, the surface area per unit area, and the trend distribution. The number of extrema measures the noisiness of the data, and the surface area measures relative amplitudes. Both are robust fractal measures that can help characterize geologic units (Gettings, 2002; Shankar, 2009). Elongation ratio is determined from the ratio of ridges and troughs to all extrema and is determined from local curvature analysis (Phillips et al., 2007). Trend distribution is a nonbiased measure of trends in the magnetic field computed from the horizontal gradient (Gettings and Bultman, 2005). The measures were computed for subgrids extracted from the magnetic anomaly grid. The subgrids were defined on the basis of geologic maps for areas of similar rock type (for example, volcanic fields) and aeromagnetic maps for areas of visually consistent texture for areas of exposed geology. These subgrids are termed candidate terranes because the rock types are known. Subgrids from areas of cover used were defined using only the aeromagnetic data, defining target terranes. Geosoft Oasis-Montaj software was used to define polygons on the maps and extract the subgrids.

In addition to the frequency distribution, a possibility function transformation was computed to produce possibility functions of the various measures normalized in the same way to facilitate unbiased comparison of the various textural measures. This transformation takes the normalized frequency and converts it to an equivalent possibility value between 0 and 1 (Dubois and Prade, 1988). These can then be used in logical combinations of the various textural measures for decision making.

For quantitative data such as survey data along profiles or as a grid a number of different methods are available. Most use the data converted to frequencies from either a histogram or some other source; that is, the number of occurrences of each variable value are normalized to a total of unity. This is analogous to a probability estimate. There are then several approaches available to calculate the possibility value. Determination of membership functions consistent with probability distribution functions of the partial uniform, exponential and standard normal classes are discussed by Civanlar and Trussell (1986). Some transformations to possibility distributions from several well known probability distributions are discussed in Dubois et al. (2004). An excellent discussion of general constraints on forward and backward transformations between probability and possibility distributions is given by Dubois et al. (1993). Transformation between probability and possibility based on evidence theory is presented by Yamada (2001) in which he compares three different transformations and concludes the method of Dubois and Prade (1983) may be the best of the three. He and Qu (2007) contrast and compare the various transformations between probability and possibility in a good summary. For data that can be expressed in a histogram, several techniques are available. Devi and Sarma (1985) show how to approximate the possibility distribution with a generalized rational function fit to the histogram data. Dubois and Prade (1983) and Dubois and Prade (1988) (also DeCETI Project, 2000) show how to calculate the possibility distribution from a normalized histogram of data. This is the method used in the present study because it depends only on the data and makes no assumptions as to functional form. If the normalized frequencies $p(u_i)$ have been sorted into descending order, then the possibility distribution is calculated as follows.

$$\pi_x(u_i) = \Pi(\{u_i\}) = i \cdot p(u_i) + \sum_{j=i+1}^{n} p(u_j) \tag{4}$$

where there are n elements u_i of the variable x, and $\sum_{j=1}^{n} p(u_j) = 1$.

Distribution of Anomaly Values Perhaps the simplest measure to compare two areas is the anomaly amplitude distribution.For the anomaly amplitude measure, the frequency distribution of the observed anomaly amplitudes observed within each subgrid was computed using statistical software written by the author.

Surface Area Distribution A second textural measure that is robust in characterizing lithologies is the distribution of amplitudes within a window of appropriate size moving over the candidate and target areas (Gettings, 2002; Shankar, 2009). In the case of profiles, the measure is the Euclidean length of the profile within the window calculated by summing the lengths of straight line segments connecting the profile data points.

$$L(x) = \sum_{i=1}^{n} \sqrt{T(x_i)^2 + (x_i)^2} \tag{5}$$

Where $T(x_i)$ is the magnetic anomaly value at x_i along the profile, and $i = 1, ., n$ ranges over the points in the window. The window is moved down the profile a specified distance and the measure $L(x)$ recalculated. This length has mixed units of the square root of anomaly squared plus distance squared. For magnetic anomaly data this is $\sqrt{(nT)^2 + (km)^2}$, where nT is nanoTesla, the unit of magnetic field strength. In the case of gridded survey data, the measure is the surface integral of the grid within the window with units of $\sqrt{(nT)^2 + (km)^2 + (km)^2}$. Noting that the surface is defined by the magnetic anomaly grid $T(x, y)$ the surface area is given by the surface integral whose surface element is the absolute value of the vector cross product of the unit vectors tangent to the surface in the x and y directions.This gives for the surface integral (Taylor, 1955):

$$S(x, y) = \int_{x_1}^{x_2} \int_{y_1}^{y_2} \left[1 + \left(\frac{\partial T(x, y)}{\partial x} \right)^2 + \left(\frac{\partial T(x, y)}{\partial y} \right)^2 \right]^{1/2} dydx \tag{6}$$

where the spatial variables x and y range over the window. Denoting the grid values of $T(x, y)$ at the ith row and j^{th} column as $T_{i,j}$, the horizontal derivatives are calculated from finite difference formulas (Young and Gregory, 1972) as follows.

$$\frac{\partial T_{i,j}}{\partial x} \doteq \frac{1}{2h_x}(T_{i+1,j} - T_{i-1,j}) \quad i\,inside\,grid \tag{7}$$

$$\frac{\partial T_{i,j}}{\partial x} \doteq \frac{-1}{2h_x}(3T_{i,j} - 4T_{i+1,j} + T_{i+2,j}) \quad i\,at\,left\,edge\,grid \tag{8}$$

$$\frac{\partial T_{i,j}}{\partial x} \doteq \frac{1}{2h_x}(3T_{i,j} - 4T_{i-1,j} + T_{i-2,j}) \quad i\,at\,right\,edge\,grid \tag{9}$$

$$\frac{\partial T_{i,j}}{\partial y} \doteq \frac{1}{2h_y}(T_{i,j+1} - T_{i,j-1}) \quad j\,inside\,grid \tag{10}$$

$$\frac{\partial T_{i,j}}{\partial y} \doteq \frac{-1}{2h_y}(3T_{i,j} - 4T_{i,j+1} + T_{i,j+2}) \quad j\,at\,bottom\,edge\,grid \tag{11}$$

$$\frac{\partial T_{i,j}}{\partial y} \doteq \frac{1}{2h_y}(3T_{i,j} - 4T_{i,j-1} + T_{i,j-2}) \quad j\,at\,top\,edge\,grid \tag{12}$$

It is assumed the grid columns (x-direction) are numbered increasing from left to right and the rows (y-direction) are numbered increasing from bottom to top. The spacing of the columns is assumed constant h_x and the spacing of the rows is assumed constant h_y for the grid. The integrand of equation 6 is accumulated in an array $s_{i,j}$ for all grid points in the (rectangular) window $i = 1, .., n_x; j = 1, .., n_y$ and integrated with the trapezoidal approximation of the double integral (Abramowitz and Stegun, 1964; Olver et al., 2011):

$$S(x, y) \doteq h_x h_y \sum_{i=1}^{n_x} \sum_{j=1}^{n_y} w_i w_j s_{i,j} \tag{13}$$

$$w_i = 1.0 \quad inside\,window; w_i = 0.5 \quad at\,left\,or\,right\,edges \tag{14}$$

$$w_j = 1.0 \quad inside\,window; w_j = 0.5 \quad at\,bottom\,or\,top\,edges \tag{15}$$

In most cases the location of the measure is taken as the center of the window in both the profile and grid (rectangular window) cases. The window is then moved a specified amount and the process repeated. In order for the measure to be comparable with other window sizes, the integral is (optionally) divided by the window area so that the resulting values are surface area per km^2. Finally, the normalized frequencies for the resulting grid of surface integral values for the area (candidate or target) is calculated and transformed to a possibility function with equation 4.

The size of the window is chosen to reflect the size of the smallest "geologic unit" of interest. This means a size which will resolve the variations characteristic of the lithostratigraphic unit as the window moves over it. For the work applied to the Arctic aeromagnetic data described below, the grid interval was 2 km and square windows 20 km in width and breadth were used. For the southern Arizona example described below, the grid interval was 50 m and windows 1 km in width and breadth were used. Gettings (2002) studied this issue and showed that the smallest "geologic unit" of interest preserved the information content, whereas larger windows lose resolution and smooth the data, thus losing information.

Curvature Measures Curvature analysis can identify extrema of a data set and provide useful information for depth estimation in potential field data (Hansen and deRidder, 2006). For profile data, Roberts (2001) and Phillips et al. (2007) give descriptions of applications to geophysical data. The application to areal data in grid form is covered in those two papers and by (Hansen and deRidder, 2006). In this case the curvature is a 2×2 tensor and these authors all calculate curvature by using a 3×3 window moving over the grid. For each window, a paraboloid is least squares fit to the nine grid values with the form:

$$f(x, y) = A + Bx + Cy + Dx^2 + Exy + Fy^2 \tag{16}$$

For a 3×3 window the coefficients have a closed solution and the coefficients D, E, and F determine the curvature tensor (matrix), which can be solved for the eigenvalues and the eigenvectors (Phillips et al., 2007). Because not every grid intersection will have a critical point within a half grid interval of the grid point at the center of the window, and to prevent duplicate solutions, the program does not put out a grid but instead outputs a "post" file, a text file that lists the critical points that have a curvature solution. For each such point, the post file contains the x, y coordinates of the solution, the row and column of the nearest grid intersection, surface value at the solution from the quadratic surface fit, the elongation factor of the quadratic fit calculated as the ratio of the eigenvalues or the depth estimate if trough or ridge, the strike direction of the fit ridge or trough in degrees clockwise from north, the two eigenvalues, and the type of feature (ridge, trough, saddle point, peak, or hole).

In order to utilize the curvature post file information to characterize texture of areas, a second program witten by the author computes four new grids from a circular moving window over the survey area that represent the total number of extrema, the elongation tendency (ratio of number of ridges and troughs to total number extrema), the average strike direction, and the variation in the strike direction estimated as the standard deviation within the window. Together with the anomaly value distribution and surface area distribution described above, this yields six textural measures characterizing areas' anomaly field. This program correctly accounts for the numerous possible cases of the circular window partially falling outside the grid boundaries and allows the user to specify the window radius. For the Arctic aeromagnetic example a 10 km radius was used, and for the Southern Arizona example a 500 m radius was used. Note that the elongation measure is distinct from the value calculated in the curvature tensor calculation because it is a ratio of all the two-dimensional (ridge or trough) solutions within the window to the total number of extrema (saddle points, peaks, holes, ridges or troughs) within the window. This measure characterizes whether a particular geologic unit tends to have elongate anomalies or equant ones. As with the other two measures, for each of the four grids of measures, the normalized frequencies are computed and transformed to possibility distributions.

Lacunarity Measure First a measure of the data as a function of scale or resolution is needed. Although several measures can be conceived, the most common is a sum of the data values in a subwindow of size $k \times k$ that moves over the $n_w \times n_w$ window advancing one node at a time. To be a measure, the function must: span the space of the data; be additive (that is, the measure over a subset that is the sum of two smaller subsets must be the sum of the measure over the two smaller subsets); and, for simple measures, it must be non-negative (Gupta (2006); Wikipedia (2012)). This last requirement means that gravity anomaly and aeromagnetic anomaly data may have to be shifted with a constant to bring all data values to a non-negative value. For example, the common definitions of area and volume are valid measures. For this work, the measure $M(k, i_{min}, j_{min})$ at scale k is defined as

$$M(k, i_{min}, j_{min}) = \sum_{i=i_{min}}^{i_{min}+k-1} \sum_{j=j_{min}}^{j_{min}+k-1} g(i, j) \qquad (17)$$

where i_{min} and j_{min} are column and row number of the lower left corner of the $k \times k$ window. The maximum number of such windows occurs if it moves one grid cell at a time. For this work, window movement is left to right along the columns of the grid until the edge is reached, then the window moves up one row and advances left to right along the columns again. For the one grid cell movement, the number of windows in which a measure at scale k can be calculated is

$$N(k) = (L/\delta - k + 1)^E = (L/\delta - k + 1)^2 \tag{18}$$

$$L = n_c\delta = n_r\delta, \quad n_c = n_r \tag{19}$$

where E, the Euclidian dimension is 2. L is the dimension of the area the measure is calculated over, $L = n_w\delta$ here for the $n_w \times n_w$ window. Figure 1 illustrates this scenario for the case of scale $k = 3$. The scale k varies from 1 to n_w in steps of 1.

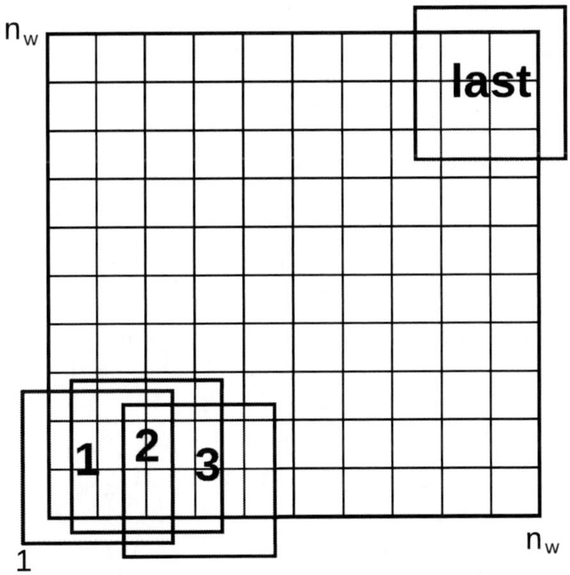

Figure 1. Diagram showing a grid window of n_w columns and n_w rows, and how a 3×3 window enclosing 9 grid intersections, for example, moves across the grid one column or row at a time. Windows labelled “2” and “3” have been displaced slightly up and down to show the windows clearly. The window moves across all the columns, then moves up a row and across the columns again, until the last window labelled “last.”

Let $n(M, k)$ be the number of windows of scale k with measure M. The probability $Q(M, k)$ for a given value of the measure M and scale k is then

$$Q(M, k) = n(M, k)/N(k) \tag{20}$$

The statistical moments of the measure distribution at scale k are then

$$Z_Q^{(q)}(k) = \sum_M M^q Q(M, k) \tag{21}$$

$$= \frac{1}{N(k)} \sum_M M^q n(M, k) \tag{22}$$

$$= \frac{1}{N(k)} \sum_{i_{min}} \sum_{j_{min}} M^q(k, i_{min}, j_{min}) \tag{23}$$

where the last right hand side results because the sum is over all measure values M. The lacunarity $\Lambda(k)$is then defined for scale k with the first and second moments as follows (Allain and Cloitre, 1991):

$$\Lambda(k) \equiv \frac{Z_Q^{(2)}(k)}{[Z_Q^{(1)}(k)]^2} \tag{24}$$

This yields a profile of $\Lambda(k)$ vs k for the $n_w \times n_w$ window. Computation of these profiles for the $n_w \times n_w$ windows covering the data grid $g(i, j)$ provides the lacunarity measure as a series of lacunarity plots, one for each $n_w \times n_w$ window.

The interpretation of the lacunarity measure is that it describes the distribution of gaps or voids of varying sizes in the data (Karperien (2004); Cheng (1997)). It has been described with words such as "patchiness," "pittedness" and "clustering" (Karperien (2004); Feder (1988)). Lacunarity quantifies the heterogeneity or translational invariance of the data (Plotnick et al., 1993). In general, small lacunarity implies a more nearly homogeneous distribution of gaps or clusters (gaps or clusters nearly the same size), whereas a larger value implies a more heterogeneous distribution (gaps or clusters with many different sizes). These statements are easier to visualize in the case of binary data where a grid cell is either zero or non-zero so the zeroes are the voids or gaps. In the case of lacunarity for continuous data such as an aeromagnetic anomaly grid, the meaning is less clear. In this case the lacunarity measure describes the distribution of clusters of data values that are similar, somewhat analogous to a histogram.

Multifractal Spectrum Measure The statistical moments of the measure M are given by

$$Z^{(q)}(k\delta) = \mathbf{E}\{M^q\} \doteq \frac{1}{N(k)} \sum_{i_{min}} \sum_{j_{min}} M^q(k, i_{min}, j_{min}) \tag{25}$$

where $\mathbf{E}\{M^q\}$ is the expectation value and it is noted that the spacing δ does not affect the calculation because the grid values are not an explicit function of δ. The partition function (Muzy et al. (1993); Cheng and Agterberg (1995); Cheng and Agterberg (1996)), which describes how the measure is distributed within the set, can be written as

$$X_q(k\delta) \approx (\frac{L}{k\delta})^E \mathbf{E}\{M^q\} \tag{26}$$

$$= (\frac{L}{k\delta})^E Z^{(q)}(k\delta) \tag{27}$$

$$= (\frac{L}{k\delta})^E \frac{1}{N(k)} \sum_{i_{min}} \sum_{j_{min}} M^q(k, i_{min}, j_{min}) \tag{28}$$

If the measure M is multifractal, the partition function satisfies a power law relation for the scale $k\delta$ with exponents a function of q (Cheng, 1997)

$$X_q(k\delta) \propto (k\delta)^{\tau(q)} \tag{29}$$

$$= (\frac{L}{k\delta})^E Z^{(q)}(k\delta) \tag{30}$$

Then

$$(k\delta)^E (k\delta)^{\tau(q)} \propto L^E Z^{(q)}(k\delta) \propto Z^{(q)}(k\delta), L^E = L^2 constant \tag{31}$$

and

$$Z^{(q)}(k\delta) \propto (k\delta)^{\tau(q)+E} = (k\delta)^{\tau^*(q)} \tag{32}$$

where $\tau^*(q) = \tau(q) + E$. Thus the moments of the measure can be used to determine the function $\tau^*(q)$ for a range of q. This is accomplished by fitting a least squares line to $ln(Z^{(q)}(k\delta))$ as a function of $ln(k\delta)$ for each moment order q, and the slope of the resulting fitted line is an estimate of $\tau^*(q)$. Once $\tau^*(q)$ as a function of q is obtained, the function $f(\alpha)$, α is obtained using (Feder (1988); Cheng (1997))

$$\alpha^*(q) = \frac{d\tau^*(q)}{dq} = \alpha(q) \tag{33}$$

$$f^*(\alpha) = \alpha^*(q)q - \tau^*(q) = f(\alpha) - E \tag{34}$$

The derivative in equation 33 is usually determined numerically using the central difference approximation (Abramowitz and Stegun, 1964)

$$\frac{d\tau^*(q)}{dq} \doteq \frac{\tau^*(q_{q+1}) - \tau^*(q_{q-1})}{q_{q+1} - q_{q-1}} \tag{35}$$

where the subscripts $q + 1$ and $q - 1$ refer to the $\tau^*(q)$ and q values ahead and behind the q value estimated since q may be non-integer. If $\tau(1) \neq 0$ or $\tau^*(1) \neq E$, then use

$$\tau^*(q) = \tau(q) + \tau^*(1) \tag{36}$$

$$f^*(\alpha) = f(\alpha) - \tau^*(1) \tag{37}$$

For a particular multifractal, the spectrum $f(\alpha)$ versus α plots as an inverted "U" shape. The plot will normally have a range between the exponents α_{min} and α_{max}. The maximum of $f(\alpha)$ occurs at α_s, the support dimension (Feder, 1988), which for continuous fields will be the Euclidean dimension (2 for the two-dimensional grids considered here). The plot is not generally symmetric about α_s, and the spectrum to the left ($\alpha_{min} \leq \alpha \leq \alpha_s$) of $f(\alpha)_{max}$ is determined by data at larger scales (resolutions). The spectrum to the right ($\alpha_s \leq \alpha \leq \alpha_{max}$) of $f(\alpha)_{max}$ is determined by data at smaller scales (resolutions). Because a grid is a low pass filter at the grid spacing δ and because data at very small spacings are rarely available in survey datasets, the right side of the spectrum is often poorly defined. Thus the left-hand of the $f(\alpha)$ curve is the most definitive in using the multifractal spectrum to estimate lithology.

The multifractal spectrum is based upon the scaling of the expectation values of the moments of the measure (equation 25). The expectation is an averaging operator and so different distributions in space can have the same measure distribution and thus the same or similar multifractal spectrum. Although the lacunarity is a function of the same first and second moments of the measure used in determining the multifractal spectrum, it often distinguishes between differing distributions with the same multifractal spectrum (Karperien, 2004). Thus the use of both measures should better discriminate between differing lithologies than either textural measure by itself.

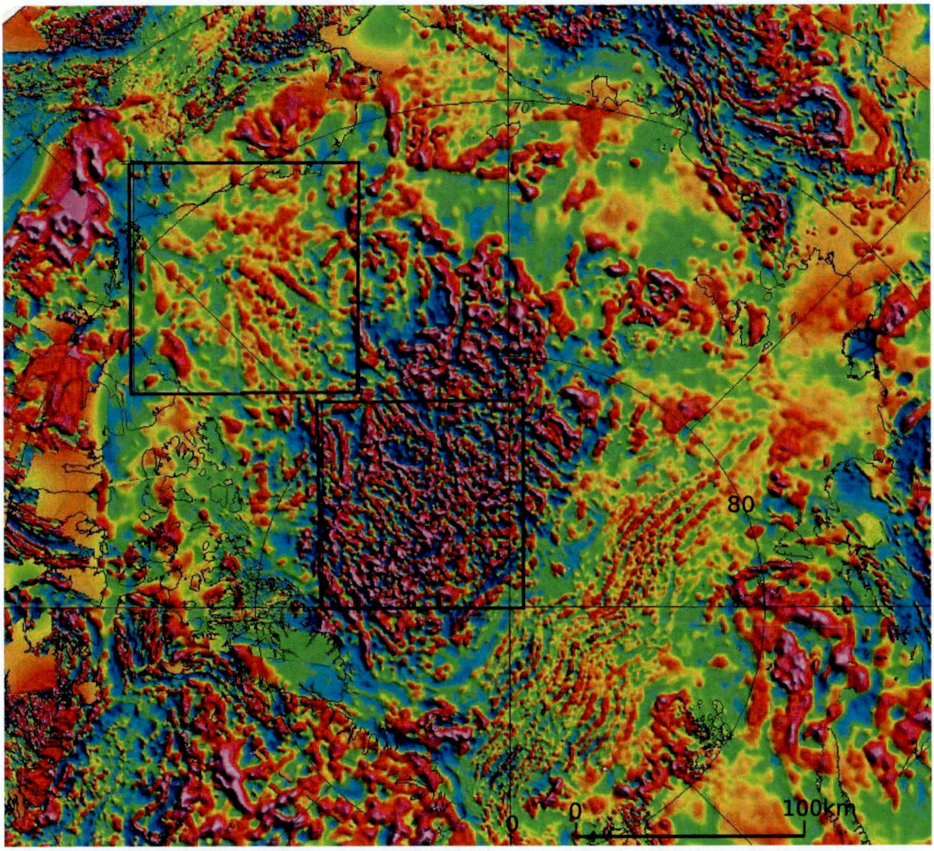

Figure 2. Aeromagnetic map of the Arctic Sea area, 2km grid interval (Gaina et al., 2010). Dark rectangles outline the two subareas characterized here. The central area is designated HAMH and the rectangle in the upper left is designated CANBASIN. The vertical line at the bottom labeled "0" is zero degrees longitude and the circle labeled "80" is 80 degrees north latitude. The range of magnetic anomalies is about 225 nT from the most red to the most blue colors in the shaded relief; note however that the total range of anomalies is about 1500 nT as shown in Figure 3.

4. Example Applications to Aeromagnetic Data

4.1. Quantifying Textures of Two Different Map Areas

The first example presented here used an aeromagnetic map of the Arctic Ocean and surrounding areas (Gaina et al., 2010), illustrated in Figure 2. The aeromagnetic map and gridded data compilation were fully described and released in Gaina et al. (2010) on dvd media and online in Gaina et al. (2011). The aeromagnetic and gravity maps were a part of the international Circum-Arctic Mapping Project to compile and interpret Arctic earth science, with particular emphasis on petroleum geology (Spencer et al., 2011). As part of this project (Saltus et al., 2011) completed a regional geodynamic interpretation of the aeromagnetic map in which they qualitatively identified 57 magnetic domains correlated with tectonic domains. A preliminary version of the textural measures (excluding lacunarity and multifractal spectrum) as an extension to Oasis/Montaj had been completed by 2016, and Saltus was experimenting with the extension applied to some of the domains from the Arctic aeromagnetic study, but the results seemed incorrect. Saltus consulted with the author, and it was decided to do two regions that were clearly distinct using the author's software to provide a test case for comparison with the Oasis/Montaj extension results. Two regions consisting of subgrids at a grid interval of 2 km were chosen on the basis of visually distinct texture (Figure 2) and textural measures were computed in each subgrid to quantify the textural differences between the two regions. This provides a set of quantitative measures that can be studied to show how well the measures discriminate in the case of two areas of clearly different anomaly distributions. The central subgrid, termed HAMH, has 444 columns and 457 rows (n=202,908), a minimum of -764 nT, maximum of 1367 nT, a mean of 47 nT and a standard deviation of 225 nT. The upper left subgrid, termed CANBASIN, has 497 columns and 511 rows (n=253,967), a minimum of -304 nT, maximum of 552 nT, a mean of -19 nT and a standard deviation of 64 nT. Area CANBASIN appears to have a more persistent ridge and trough texture with longer horizontal anomaly extents than area HAMH, which has more areally restricted anomalies although they are of larger amplitude.

The eight textural measures defined above were computed for both subgrids and compared. For brevity, only the frequency distribution and the possibility distribution are shown for the anomaly amplitudes in Figures 3 and 4. For the rest of the measures, except lacunarity and multifractal spectrum, only the possibility distribution is shown as they contain all the information and are more easily compared between terranes. The distribution of anomaly amplitudes shows a narrow, approximately symmetric distribution for CANBASIN, whereas area HAMH has a broad distribution with a long tail to larger anomaly values (Figures 3 and 4). The number of extrema in a 10 km radius circular window moving 10 km (5 grid units) at a time over area CANBASIN shows fewer extrema in

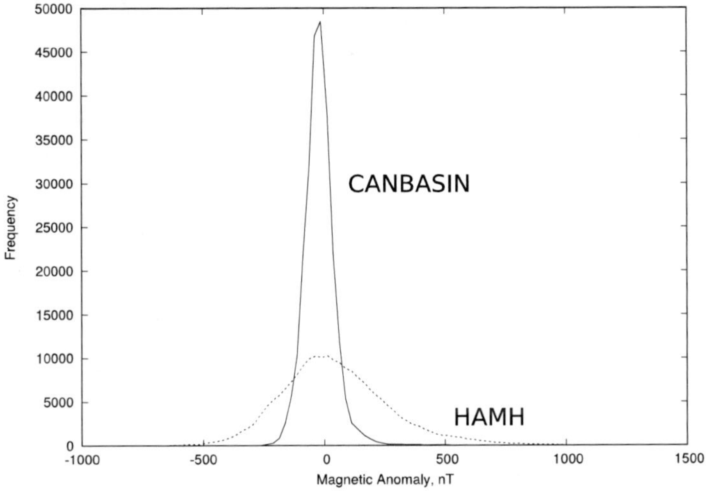

Figure 3. Frequency histogram of the aeromagnetic anomaly amplitudes for each of the two areas. Solid line is for area CANBASIN and dashed line is for area HAMH.

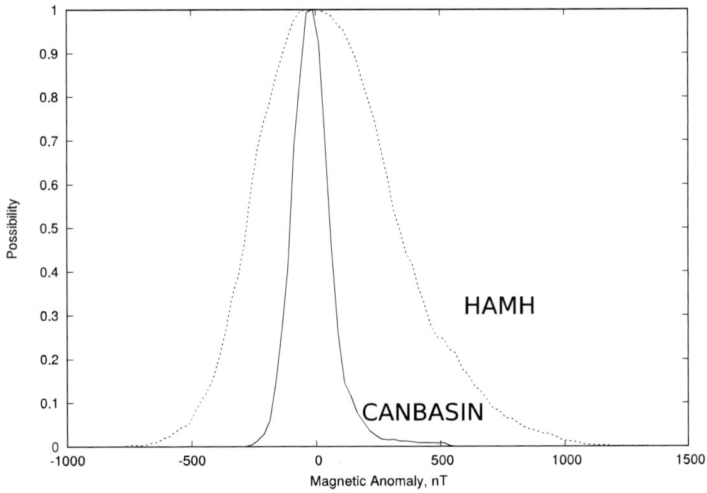

Figure 4. Possibility distribution of the aeromagnetic anomaly amplitudes for each of the two areas. Solid line is for area CANBASIN and dashed line is for area HAMH.

the window and has a long tail to larger values (Figure 5). HAMH is more symmetrically distributed about approximately twice the number of extrema in the window from area CANBASIN.

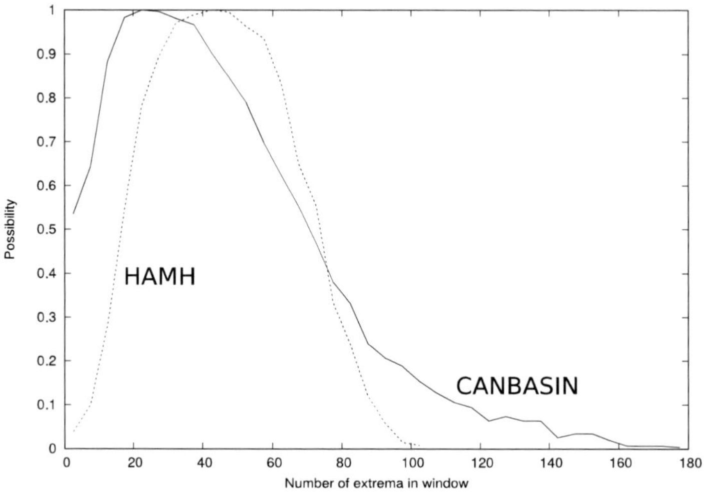

Figure 5. Possibility distribution of the number of extrema in a 10 km radius window moving over each of the two areas. Solid line is for area CANBASIN and dashed line is for area HAMH.

The elongation ratio for the two areas is shown in Figure 6. As is apparent from the aeromagnetic maps, both have many elongate anomalies, but area CANBASIN has many more with a ratio of 0.95 or more and its distribution is much wider with a long tail to smaller ratios and some anomalies nearly circular compared to area HAMH. The mean strike directions for the two areas are shown in Figure 7. Note that "north" here refers to the top of the map area rather than true north as the area includes the north geographic pole. Area CANBASIN contains mostly north-northwest trending anomalies whereas area HAMH is characterized by more northerly trends.The distribution of strikes is significantly broader in the case of area CANBASIN. The standard deviation of the strike directions within the windows is shown in Figure 8. Area HAMH has a much narrower distribution and smaller variation than area CANBASIN which has larger variability on average and a much broader distribution.

The anomaly surface area distribution in a rectangular 20 km by 20 km window moving 2 km (1 grid unit) at a time is shown in Figure 9. The areas

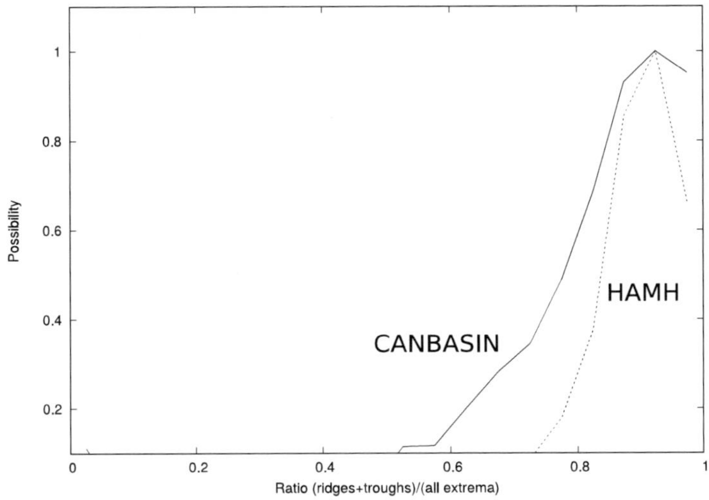

Figure 6. Possibility distribution of the elongation ratio in a 10 km radius window moving over each of the two areas. Solid line is for area CANBASIN and dashed line is for area HAMH.

of anomalies within the window are significantly smaller and tightly grouped for area CANBASIN and much larger and more broadly distributed for area HAMH. Both distributions are long tailed to larger area values.

The lacunarity measure for each area is shown in Figure 10. Although the two curves appear similar and close to each other, note that both scales are logarithmic and so there is significant separation on the abscissa between the two areas and area HAMH achieves a given lacunarity at a somewhat smaller box-size (scale) than area CANBASIN. Note the lacunarity value does not change strongly until $ln(boxsize) = 5.5$ which is approximately a 240×240 point window. This is in part due to the smoothness of the magnetic field at shorter scales and in part due to the low-pass filter effect of the grid generation operation.

The multifractal spectra for both areas is shown in Figure 11. Due to computational considerations, the software is limited to a maximum window size of 300×300 grid points. Here, windows of 251×251 grid points (500×500 km) were used. As discussed above, the left-hand side of the spectrum is most meaningful for most geophysical work as it is most controlled by the larger horizontal scales. Two of the spectra for CANBASIN are close to the spectra for HAMH on the left-hand side and the other three have a smaller range of the

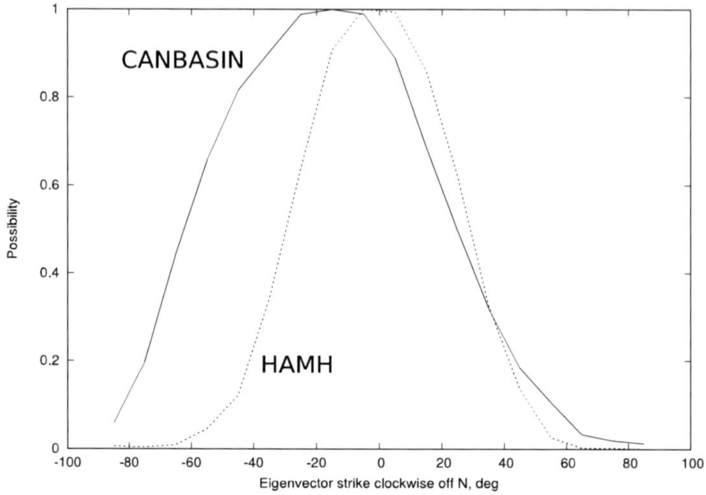

Figure 7. Possibility distribution of the mean strike distribution in a 10 km radius window moving over each of the two areas. Solid line is for area CANBASIN and dashed line is for area HAMH.

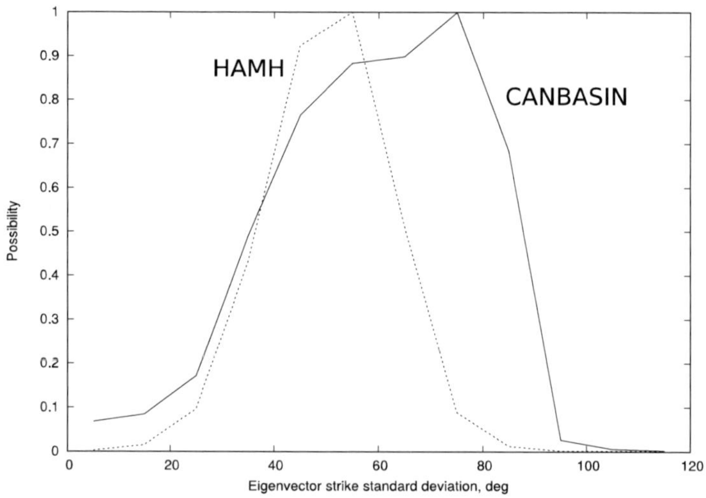

Figure 8. Possibility distribution of the standard deviation of strike distribution in a 10 km radius window moving over each of the two areas. Solid line is for area CANBASIN and dashed line is for area HAMH.

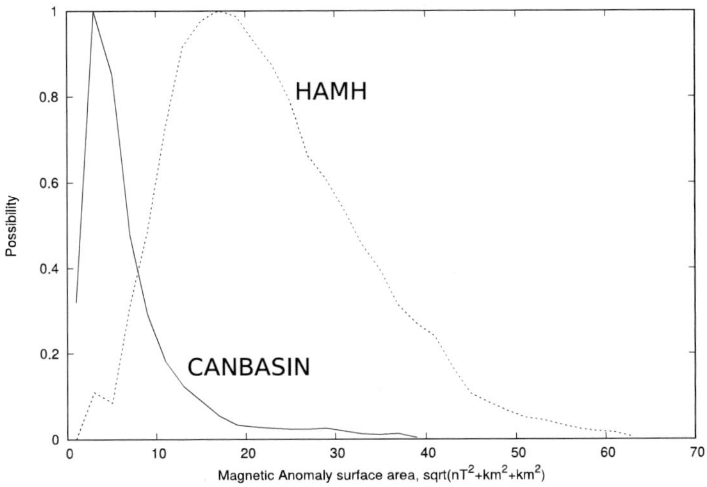

Figure 9. Possibility distribution of the aeromagnetic anomaly surface area distribution in a 20 km x 20 km window moving over each of the two areas. Solid line is for area HAMH and dashed line is for area CANBASIN.

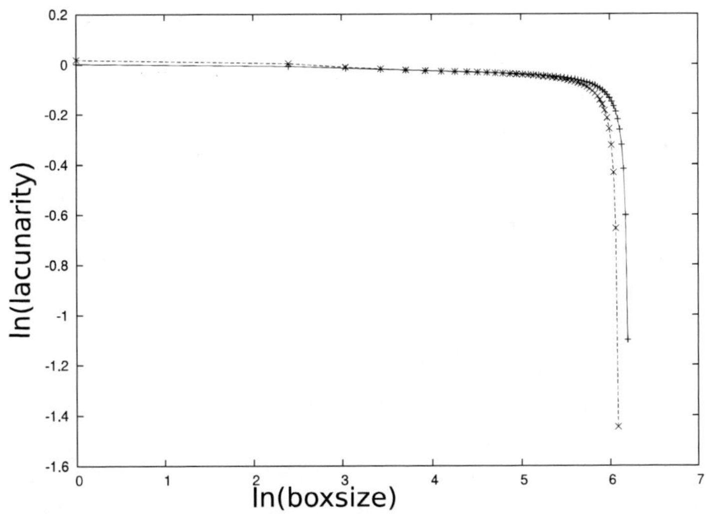

Figure 10. Lacunarity measure of the aeromagnetic anomaly for the two areas. Plus symbols are for the CANBASIN area and x symbols are for the HAMH area.

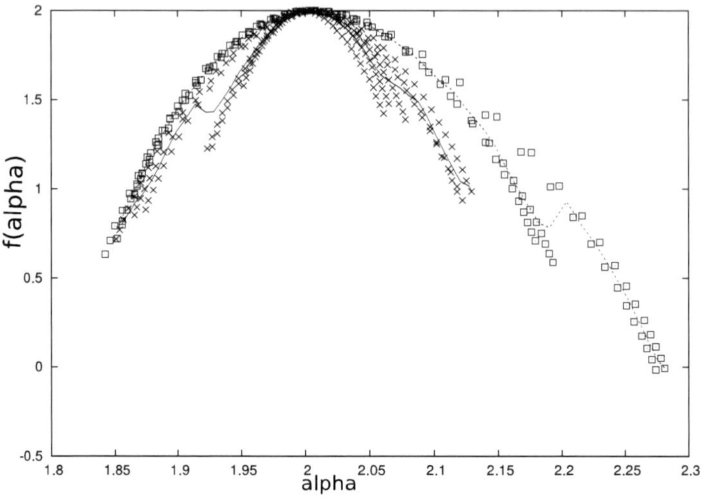

Figure 11. Multifractal spectrum of the aeromagnetic anomaly in a 500 km x 500 km window moving over each of the two areas. Solid line is the moving window centroid of the spectra for the CANBASIN area windows and the dashed line is the moving window centroid for the HAMH area. Symbols with crosses are the window spectra for the CANBASIN area and boxes are spectra for the HAMH area. See text for details.

scaling exponent α. All of the CANBASIN spectra have a narrower range of α than HAMH, which are very consistent on the left-hand side, corresponding to a more homogeneous distribution of anomaly shapes than the CANBASIN anomalies. In order to produce some sort of mean or expectation curve for the multiple spectra for each area, the centroid of the spectra were computed in a moving window and then all the spectra were plotted. The solid line is the result for CANBASIN and the dashed line in the figure is the result for area HAMH, using a window $1/20$ of the α range moving $1/2$ the window width at a time. Even taking into account the dispersion, the two lines are distinct (Figure 11).

Each of the measures provided a distinctive distribution between the two areas, and characterizes a physical property of the spatial distribution of potential field amplitudes in a quantified way. The possibility transformation is convenient as it provides a standardized way to compare the measures and allows the quantitative combination of measures unbiased by varying ranges. Logical combinations of measures using and (intersection) and or (union) are also pos-

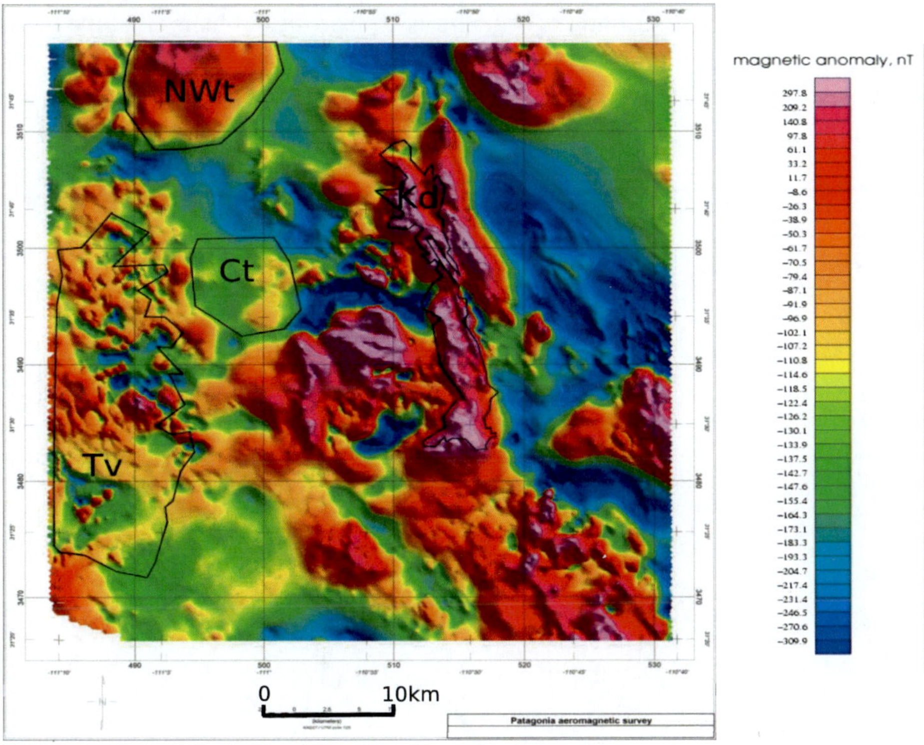

Figure 13. Aeromagnetic map (50 m grid interval) of the Patagonia area, southeast Arizona (Phillips, 2002) showing the two covered areas (Ct and NWt) whose bedrock lithology was estimated and the candidate terranes Tv and Kd.

rectangular window or a circular window with a radius of 500 m, moving 1 grid interval at a time (Gettings, 2009). Computed measures for the target areas are plotted with the candidate terrane which most closely matched the distribution. It should be noted that for the covered target areas discrimination of the smaller variations in the measures will not be as good as for the candidate terrane because smaller variations are within the noise envelope. This is particularly true for aeromagnetic anomalies compared to gravity anomalies as the amplitude decreases as the cube of the distance for magnetic sources as opposed to the square of the distance for gravitational sources. Originally quantitative logical combinations of the possibility functions were planned (AND, OR, NOT, etc) but were precluded by project funding and time limitations. However in this case, visual examination of the distributions of textural measures establishes that the Ct target is most likely Tv and the NWt target is most likely an intrusive from the Kd suite.

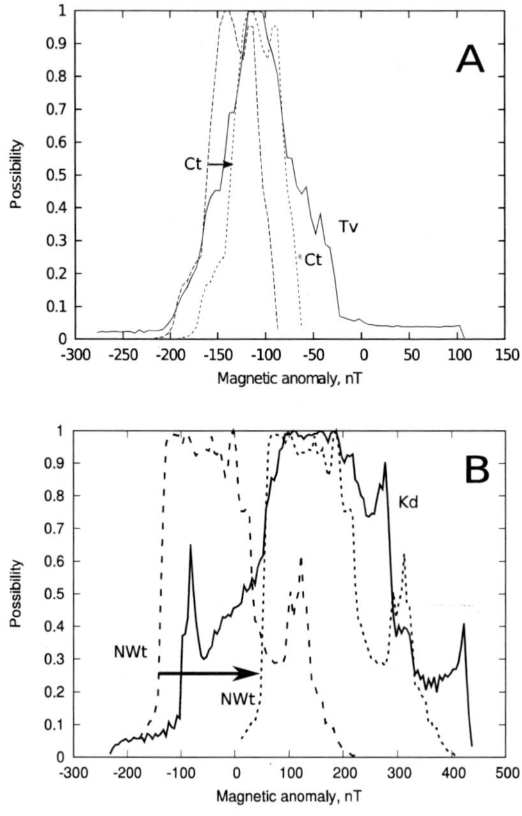

Figure 14. Possibility distributions of magnetic anomaly amplitude. A. Amplitude distribution for the Ct and Tv areas, solid line is the Tv area, dashed line is Ct area and short dashed line is the Ct distribution shifted to best fit with Tv (minimum Chi-square) to remove regional anomaly variation. B. Amplitude distribution for the NWt and Kd areas, solid line is the Kd area, dashed line is NWt area and short dashed line is the NWt distribution shifted to best fit (minimum Chi-square) with Kd to remove regional anomaly variation.

Figure 14 compares the anomaly amplitude distributions for the candidate terranes and the target areas. Magnetic and gravity fields are superpositions of

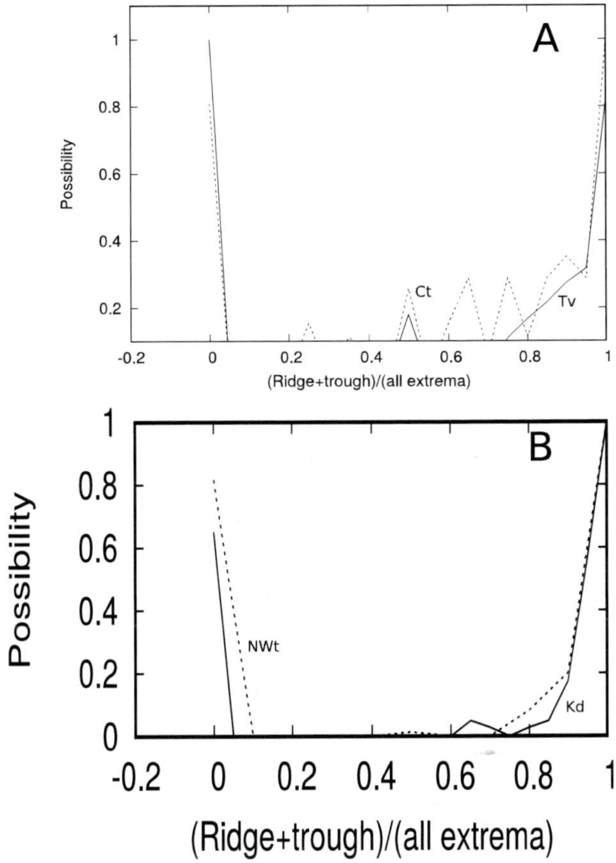

Figure 15. Possibility distributions of anomaly elongation. A. Elongation ratio for the Tv (solid line) and Ct (dashed line) areas. B. Elongation ratio for the NWt (dashed line) and Kd (solid line) areas.

the fields of all the contributing sources. Consequently as one moves about a survey area contributions due to deeper sources may offset the distribution to larger or smaller values. Therefore in order to properly compare distributions of anomaly amplitude one distribution may need to be shifted along the anomaly amplitude axis relative to the other. This has been done for the two target areas in Figure 14 by computing the Chi-squared measure of goodness-of-fit for various shifts of one distribution relative to the other and choosing the offset that minimized the Chi-squared measure. It is clear that target Ct is most similar to candidate terrane Tv and target NWt is most similar to candidate terrane Kd in the distribution of anomaly amplitudes (Figure 14).

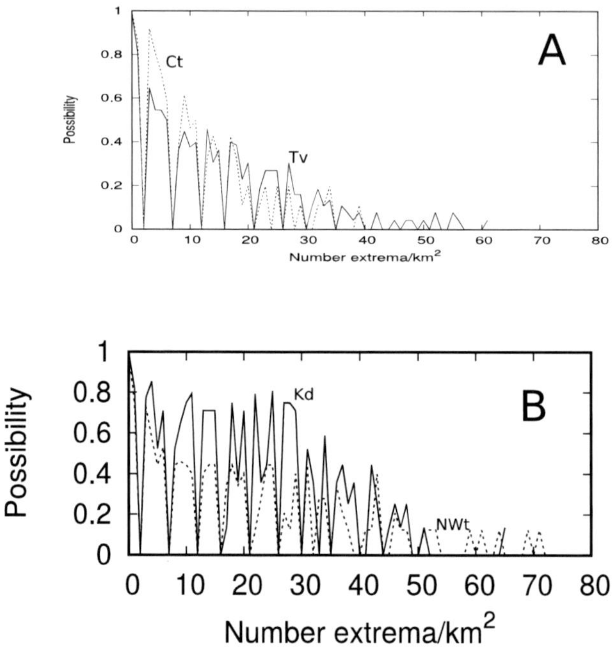

Figure 16. Possibility distributions of number of extrema. A. Number of extrema for the Tv (solid line) and Ct (dashed line) areas. B. Number of extrema for the NWt (dashed line) and Kd (solid line) areas.

Figure 15 plots the elongation ratio for the four areas and the Ct and Tv distributions are distinct in containing ratios in the 0.1-0.6 range that do not occur in the distributions for NWt and Kd. Figure 16 shows possibility distributions of the number of extrema for the four areas. The measure decreases uniformly for the Ct and Tv areas but persists at a nearly uniform level out to about 35 per square km for the NWt and Kd areas.

The most similar measure distributions are those for strikes (Figure 17). This is due to extensional basin and range deformation that at least in part post-dates the formation of the all the rock types of interest. Although the distributions differ in detail, especially amplitude, they are visually similar and not as discriminating as the other measures. The variation of strikes shown as the standard deviation of strikes within the moving window is more discriminating

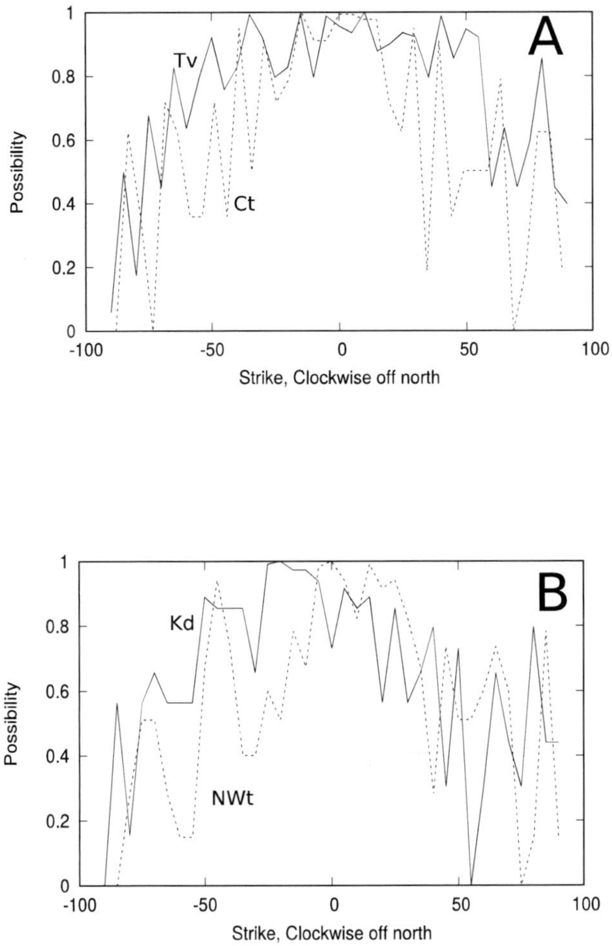

Figure 17. Possibility distributions of strikes. A. Strikes for the Tv (solid line) and Ctv (dashed line) areas. B. Strikes for the NWt (dashed line) and Kd (solid line) areas.

(Figure 18). The distribution for areas Ct and Tv is more restricted in a broad peak centered on about 50 degrees. The same measure for areas NWt and Kd is essentially uniformly distributed over the range 0-80 degrees.

Finally, Figure 19 shows the distribution of anomaly surface area in the moving window for the four areas. There is some noise introduced to the target

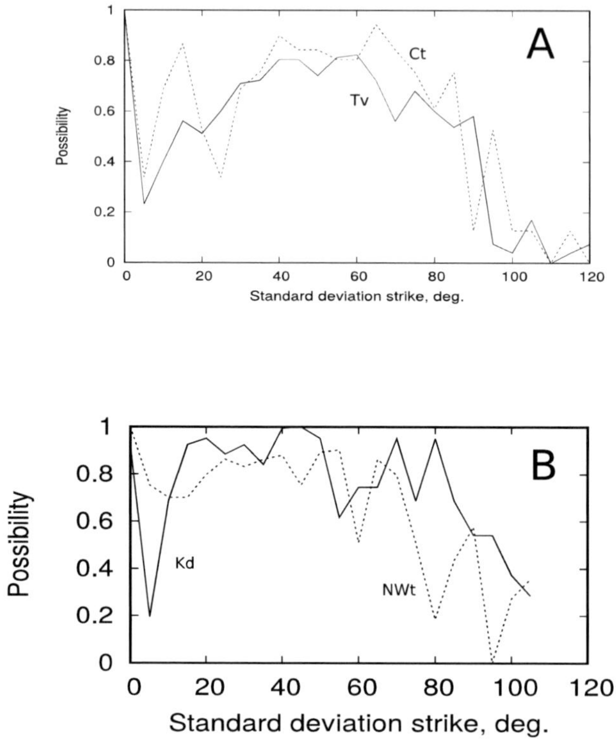

Figure 18. Possibility distributions of standard deviation of strikes. A. Strike standard deviations for the Tv (solid line) and Ct (dashed line) areas. B. Strike standard deviations for the NWt (dashed line) and Kd (solid line) areas.

signal from the small magnetization of the alluvial fill overlying the bedrock that is not present in the upward continued candidate areas. This causes a loss of signal definition at larger surface areas and is apparent in the case of both target areas Ct and NWt relative to Tv and Kd. Nevertheless the shapes are distinctive in correlating Ct with Tv and NWt with Kd, as well as the surface area values being much larger for NWt and Kd relative to Ct and Tv. Thus, based on the comparison of the six measures it is concluded that Ct bedrock is most likely Tv, and NWt bedrock is most likely Kd.

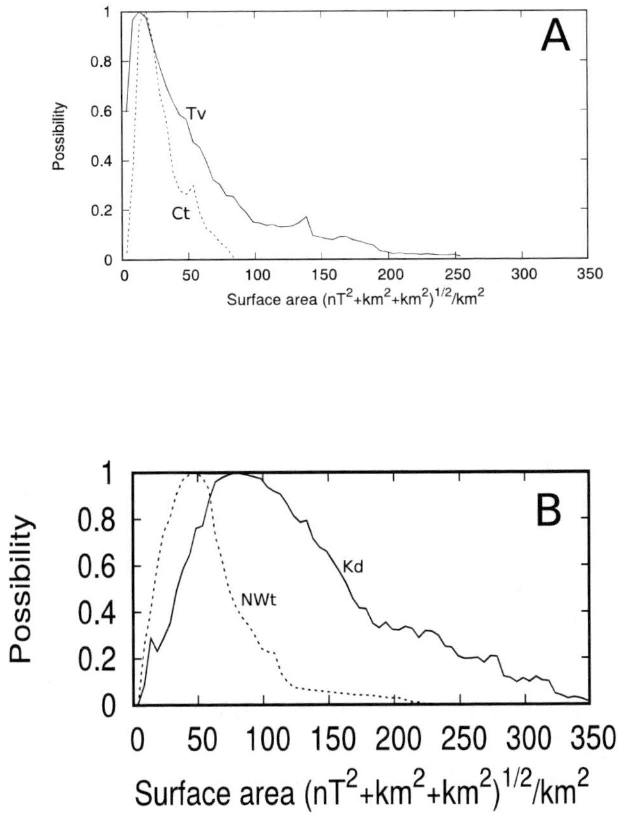

Figure 19. Possibility distributions of surface area. A. Surface area for the Ct (dashed line) and Tv (solid line) areas. B. Surface area for the NWt (dashed line) and Kd (solid line) areas.

CONCLUSION

The texture of a potential field and its textural measures have been defined as the spatial variation of potential field amplitudes and the spatial variation of measures of the amplitudes. Eight measures have been defined in terms of their distribution as measured by a window moving over the data: the amplitudes; the number of extrema; the elongation ratio; the strike of the eigenvectors; the standard deviation of the eigenvector strikes; the surface area of the amplitudes, the lacunarity measure, and the multifractal spectrum. These measures are statistically sound because they utilize large numbers of points within the windows. For the Arctic aeromagnetic example, each measure distinguished clearly between the two test areas and provided a quantitative characterization of the texture. For the southern Arizona aeromagnetic example, comparison of the basic six textural measures yields a highly likely identification of the bedrock lithology for two very different target areas beneath thick alluvial fill. Although the least discriminatory measure was strike distribution because of deformation that affected all the target and candidate areas, the strike measure could be useful to distinguish terranes that have had different deformational events with sufficiently different strikes.

Future research in this area should include application to other areas of varying scales in order to evaluate conditions under which the measures discriminate well or poorly. When there are more than two candidate terranes for a target area, proper logical combinations should result in a list ranking the likelihood that the target lithology is a particular candidate terrane. Software needs to be developed to enable one to easily make logical combinations of the possibility distributions. Finally, research into the mathematics of observed, rather than theoretical, distributions is needed. Methods of discrimination between distributions, determination of similarity or correlation, and new methods of visualizing these kinds of data are needed. Traditional measures such as mean, standard deviation, skewness, etc. are not as meaningful for these sets of observations.

The FORTRAN software used in these studies is planned to be released in the near future in the Github environment.

ACKNOWLEDGMENTS

The author thanks V. Langenheim and an anonymous reviewer for helpful reviews of the manuscript. Publication of this report was approved by the Director, U.S. Geological Survey.

REFERENCES

Abramowitz, M. and Stegun, I. A. (1964). *Handbook of Mathematical Functions with Formulas, Graphs, and Mathematical Tables*. Dover.

Allain, C. and Cloitre, M. (1991). Characterizing the lacunarity of random and deterministic fractal sets. *Physical Review A*, 44(6):3552–3558.

Blakely, R. (1996). *Potential Theory in Gravity and Magnetic Applications*. Stanford-Cambridge Program. Cambridge University Press.

Bultman, M. (2013). Investigation of the structure and lithology of bedrock concealed by basin fill using ground-based magnetic-field-profile data acquired in the san rafael basin, southeastern arizona. *U.S. Geological Survey Scientific Investigation Report 2013-5038.*, 2013:26 p.

Cheng, Q. (1997). Multifractal modeling and lacunarity analysis. *Mathematical Geology*, 29(7):919–932.

Cheng, Q. and Agterberg, F. (1995). Multifractal modeling and spatial point processes. *Mathematical Geology*, 27:831–845. 10.1007/BF02087098.

Cheng, Q. and Agterberg, F. (1996). Comparison between two types of multifractal modeling. *Mathematical Geology*, 28(8):1001–1015.

Civanlar, M. R. and Trussell, H. J. (1986). Constructing membership functions using statistical data. *Fuzzy Sets Syst.*, 18(1):1–13.

Cooper, G. (2004). The textural analysis of gravity data using co-occurrence matrices. *Computers & Geosciences*, 30(1):107 – 115.

Cooper, G. (2009). The antialiased textural analysis of aeromagnetic data. *Computers & Geosciences*, 35(3):586 – 591.

Cooper, G. and Cowan, D. (2005). The use of textural analysis to locate features in geophysical data. *Computers & Geosciences*, 31(7):882 – 890.

DeCETI Project (2000). Multi-sources information fusion for satellite image classification. electronic Report of the DeCETI Project. Contract No: GR/1996/II/0953/PI/II.1.1.c/FPC.

Dentith, M. (1995). Textural filtering of aeromagnetic data. *Exploration Geophysics*, 26:209–214.

Devi, B. and Sarma, V. (1985). Estimation of fuzzy memberships from histograms. *Information Sciences*, 35:43–59.

Dubois, D., Foulloy, L., Mauris, G., and Prade, H. (2004). Probability-possibility transformations, triangular fuzzy sets and probabilistic inequalities. *Reliable Computing*, 10:273–297.

Dubois, D. and Prade, H. (1983). Unfair coins and necessity measures: Towards a possibilistic interpretation of histograms. *Fuzzy Sets and Systems*, 10:15–20.

Dubois, D. and Prade, H. (1988). *Possibility Theory: An Approach to Computerized Processing of Uncertainty*. Plenum Press. 263p.

Dubois, D., Prade, H., and Sandri, S. (1993). On possibility/probability transformations. In *Proceedings of Fourth IFSA Conference*, pages 103–112. Kluwer Academic Publ.

Feder, J. (1988). *Fractals*. Number ISBN 0-306-42851-2. Plenum Press, New York.

Gaina, C., Werner, S., and CAMP-GMgroup (2010). Circum-arctic mapping project-gravity and magnetic maps (camp-gm). *Reports of NGU,Geological Survey of Norway, Trondheim, Norway*, (2009.010):24 p. ISSN 0800-3416.

Gaina, C., Werner, S., Saltus, R., Maus, S., Aaro, S., Damaske, D., Forsberg, R., Glebovsky, V., Johnson, K., Jonberger, J., Koren, T., Korhonen, J., Litvinova, T., Oakey, G., Olesen, O., Petrov, O., Pilkington, M., Rasmussen, T., Schreckenberger, B., and Smelror, M. (2011). Chapter 3: Circum-arctic mapping project: New magnetic and gravity anomaly maps of the arctic. *Geological Society Memoir*, 35:39–48.

Gettings, M. (2009). Identification of concealed lithologies using possibility theory and aeromagnetic data. In Williams, P. et al., editors, *Smart Science for Exploration and Mining, Proceedings of the 10th Biennial SGA Meeting of The Society for Geology Applied to Mineral Deposits, Townsville Australia, 17th - 20th August 2009, Session C4: Finding resources under cover: new geophysical and imaging techniques for exploration, Mark Gettings, ed.*, pages 796–798.

Gettings, M. and Bultman, M. (2005). A predictive penetrative fracture mapping method from regional potential field and geologic datasets, southwest colorado plateau, u.s.a. *Earth Planets Space*, 57(8):701–715.

Gettings, M. and Houser, B. (1997). Basin geology of the upper santa cruz valley, pima and santa cruz counties, southeastern arizona. *Open File Report 97-676*, page 40 p.

Gettings, M. E. (2002). An interpretation of the 1996 aeromagnetic data for the santa cruz basin, tumacacori mountains, santa rita mountains, and patagonia mountains, south-central arizona. Technical Report Open-File Report 02-99, U.S. Geological Survey.

Gettings, M. E. (2012a). Extending geologic map units beneath cover using geophysical textural measures. In *Abstract GP43B-1140 presented at 2012 Fall Meeting, AGU, San Francisco, Calif., 3-7 Dec.*

Gettings, M. E. (2012b). Multifractal model of magnetic susceptibility distributions in some igneous rocks. *Nonlinear Processes in Geophysics*, 19(6):635–642.

Gettings, M. E. (2016a). Quantitative textural measures calculation from gridded gravity and magnetic anomaly data. *Geological Society of America Annual meeting abstracts, poster 63-6.*

Gettings, M. E. (2016b). Quantitative textural measures of potential field anomaly data. *Geological Society of America 2018 Annual meeting abstracts, paper 177-7.*

Gettings, M. E. (2018). Extending proterozoic terranes beneath cover using potential fields. *Society of Economic Geologists 2018: Metals, Minerals, and Society conference, Keystone, Colorado, USA, poster.*

Gupta, M. R. (2006). A measure theory tutorial (measure theory for dummies). UWEE Technical Report UWEETR-2006-008. urldate = 2014-05-29.

Hansen, R. O. and deRidder, E. (2006). Linear feature analysis for aeromagnetic data. *Geophysics*, 71(6):L61–L67.

Haralick, R. M. (1979). Statistical and structural approaches to texture. *Proceedings of the IEEE*, 67(5):786–804.

He, L.-P. and Qu, F.-Z. (2007). A comparative study of compatibility and transformation between probability and possibility. In *Proc. Fourth International Conference on Fuzzy Systems and Knowledge Discovery FSKD 2007*, volume 1, pages 12–16.

Karperien, A. B. (2004). Defining microglial morphology: Form, function, and fractal dimension. Master's thesis, Charles Sturt University.

Muzy, J. F., Bacry, E., and Arneodo, A. (1993). Multifractal formalism for fractal signals: The structure-function approach versus the wavelet-transform modulus-maxima method. *Phys. Rev. E*, 47(2):875–884.

Olver, F. W. J., Lozier, D. W., Boisvert, R. F., and Clark, C. W. (2011). Digital library of mathematical functions (dlmf).

Oxford English Dictionary (1971). *Compact Edition of the Oxford English Dictionary.* Oxford University Press.

Peterson, J. A., Berquist, J. R., Reynolds, S. J., Page-Nedell, S. S., Oland, G. P., and Hirschberg, D. M. (2001). Geologic map of the tucson and nogales quadrangles (arizona, scale 1:250,000): A digital database. *U. S. Geological Survey Open-File Report 01-275*, page 25 pp.

Phillips, J., Hansen, R., and Blakely, R. (2007). The use of curvature in potential-field interpretation. *Exploration Geophysics*, 38:111–119.

Phillips, J. D. (2002). Processing and interpretation of aeromagnetic data for the santa cruz basin - patagonia mountains area, south-central arizona. *U.S. Geological Survey Open-File Report 2002-98.*

Plotnick, R., Gardner, R., and O'Neill, R. (1993). Lacunarity indices as measures of landscape texture. *Landscape Ecology*, 8(3):201–211.

Raines, G. L. (2002). Description and comparison of geologic maps with fragstats a spatial statistics program. *Computers & Geosciences*, 28(2):169 – 177.

Rao, A. (1990). *A Taxonomy for Texture Description and Identification.* Springer-Verlag Inc., New York.

Rehrig, W. A. and Reynolds, S. J. (1980). *Cordilleran Metamorphic Core Complexes*, volume Memoir 153, chapter Geologic and geochronologic reconnaissance of a northwest-trending zone of metamorphic core complexes in southern and western Arizona, pages 131–157. Geologic Society of America.

Renno, C. D., Freitas, C., and SantAnna, S. J. S. (1998). A system for region image classification based on textural measures. *European Space Agency-Publications-Esa SP*, 434:159–164.

Roberts, A. (2001). Curvature attributes and their application to 3d interpreted horizons. *First Break*, 19(2):85–100.

Saltus, R. W., Miller, E. L., Gaina, C., and Brown, P. J. (2011). Chapter 4 regional magnetic domains of the circum-arctic: a framework for geodynamic interpretation. *Geological Society, London, Memoirs*, 35(1):49–60.

Shankar, V. (2009). Texture-based automated lithological classification using aeromagenetic anomaly images. *U.S. Geological Survey Open-File Report 2009-1206*, page 130 p.

Shapiro, L. G. and Stockman, G. C. (2001). *Computer vision*. Prentice Hall, Upper Saddle River, NJ.

Spencer, A. M., Embry, A. F., Gautier, D. L., Stoupakova, A. V., and Soerensen, K., editors (2011). *Arctic Petroleum Geology*, volume 35. Geological Society of London.

Taylor, A. E. (1955). *Advanced Calculus*. Blaisdell Publishing Company, a division of Ginn and Company.

Tuceyran, M. and Jain, A. (1998). Texture analysis. In Chen, C., Pau, L., and Wang, P., editors, *The Handbook of Pattern Recognition and Computer Vision, 2 nd ed.*, pages 207–248. World Scientific Publishing Co., Hackensack, New Jersey.

Weszka, J. S., Dyer, C. R., and Rosenfeld, A. (1976). A comparative study of texture measures for terrain classification. *Systems, Man and Cybernetics, IEEE Transactions on*, SMC-6(4):269–285.

Wikipedia (2012). Measure (mathematics) — wikipedia, the free encyclopedia. *Wikipedia, the free encyclopedia*. [Online; accessed 26-September-2012].

Yamada, K. (2001). Probability-possibility transformation based on evidence theory. In *Proc. Joint 9th IFSA World Congress and 20th NAFIPS International Conference*, volume 1, pages 70–75.

Young, D. M. and Gregory, R. T. (1972). *A Survey of Numerical Mathematics, volume I*. Addison-Wesley Publishing Company.

In: An Introduction to Approaches … ISBN: 978-1-53618-680-2
Editor: Yi-Tung Chan

Chapter 7

MEDICAL APPLICATIONS OF ENSEMBLE LEARNING

Ciro Comparetto[1,*] and Franco Borruto[2]
[1]Obstetrics and Gynecology, Florence, Italy
[2] Obstetrics and Gynecology, Consultant in Health Policy of the Government of Monaco, Principality of Monaco

ABSTRACT

Ensemble learning (EL) in statistics and machine learning (ML) is a series of methods that use multiple models to obtain better predictive performance than the models they are made of. Unlike the statistical set that is considered infinite, in statistical mechanics, in an ML set, we refer only to particular and finite set of alternative models. In other words, EL is an ML technique based on the construction of multiple hypotheses. The term ensemble means "collection" and indicates a set of assumptions that the machine elaborates. Unlike simpler systems, to formulate a prediction, the computer does not extrapolate a single theory but many. In ML, a hypothesis is a decision tree (DT). In recent years, due to the growing computational power that allows ample substantial-considerable EL training in a reasonable amount of time, the number of its applications has increasingly grown. Some of the forms of ensemble classifiers are: 1) remote sensing: a) land cover mapping; and b) change detection;

2) computer security: a) distributed denial of service; b) malware detection; and c) intrusion detection; 3) face recognition; 4) emotion recognition; 5) fraud detection; 6) financial decision-making; and 7) Medicine.

In the medical field, ensemble classifiers have been successfully applied in neuroscience, proteomics, and medical diagnosis, as in neurocognitive disorders (e.g., Alzheimer's disease or myotonic dystrophy) detection based on magnetic resonance imaging (MRI) data sets. Therefore, there has been enormous attention in the area of intersection between artificial intelligence (AI) and Medicine. In general, it can be observed that the leading experts predict a heavy influence on world health by the new AI and ML technologies between 2025 and 2030. This time frame coincides with many other forecasts that indicate how the use of digital intelligence will affect many different fields of culture and human activities (public administration, economics, management, industrial supply chain, transport, energy management, and more). At the moment, in the western world, there are about 50 companies actively involved in the development of ML systems aimed specifically at healthcare uses. In addition to these established companies, we also have several start-ups and university spinoffs that, almost daily, turn their attention to the world of Medicine. Not to mention the realities of the eastern (China, Japan, and Taiwan) and the Arab world.

When it comes to the effectiveness of the use of ML in Medicine, it is always said that more data almost always produces better results, and the health sector is sitting on a gold mine of data that has so far been scarcely exploited. In concrete terms, the main innovations that AI would bring to Medicine are at least 10: 1) better application and definition of diagnostic criteria for various diseases; 2) possibility of applying precision medicine in the real world; 3) automation of prototyping and synthesis of new drugs; 4) automation of reporting in pathological anatomy; 5) automation of reporting in radiology and radiotherapy; 6) chatbots and vocal assistants in the relationship with the public, in educational and psychoeducational interventions, and patient support; 7) optimization of randomized clinical trials (RCT) in the study of new treatments; 8) automation of data collection or electronic health record (EHR) and study of health big data; 9) prediction of epidemic events, and 10) prediction and study of psychic and behavioral phenomena. Each of these elements of innovation has within it several subsets of different characteristics relating to the various medical specialties. For example, psychiatry: in the immediate future, thanks to ML, we will talk about precision psychiatry, the use of advanced imaging techniques for diagnosis and therapy, the targeted development of new drugs, diagnostic objectification, expert systems to choose the best therapeutic interventions, and so on.

Keywords: ensemble learning, machine learning, deep learning, artificial intelligence, healthcare, big data, medicine, obstetrics, gynecology

INTRODUCTION

Ensemble learning (EL) is a machine learning (ML) technique based on the construction of multiple hypotheses. The term ensemble means "collection" and indicates a set of hypotheses that the machine elaborates. Unlike simpler systems, to formulate a prediction, the machine does not extrapolate a single hypothesis but many. In ML, a hypothesis is a decision tree (DT). A DT is constructed by selecting first the attributes considered most relevant, then all the others according to a hierarchical classification. Each DT has a forward margin of error. By building multiple DT from the same dataset, the probability of error is reduced, because each tree follows a different method of reasoning. Therefore, trees are likely to provide different answers. To make a decision, the machine consults all the DT. It has extrapolated. Then it chooses the answer provided by the majority of DT.

The EL process is divided into two phases:

1) the construction of hypotheses: the machine uses a training set (training data), and from this data set, it builds a DT. In the case of EL, the machine processes *N* different DT. There are several techniques for building multiple DT, starting from the same training set. For example, the boosting algorithm. Each DT differs from the others in shape because the machine selects the attributes differently. It is a hypothesis and a method of reasoning in itself. It is therefore probable that, under the same conditions, each tree will provide a different answer to solve practical problems; and
2) the decision phase: to solve a problem, the machine consults all the DT of the hypothesis space it has built. DT provides different answers, because they follow different logics. Each tree has a *px* probability of error. While providing different answers, some DT

are likely to return the same answer. At this point, the machine chooses the most frequent solution or the one provided by the majority of DT. This reduces the likelihood of error in the decision-making system.

Depending on the circumstances, the decision rule may vary. For example, if the cost of the error is very high, the machine adopts the solution provided by 70-80% of DT. If, on the other hand, the cost is low, it selects the most frequent solution, even if a few DT provides it. When using a single training set, DT is not truly independent of each other. If the training set contains misleading data, it is likely to affect in the construction of all DT. A training set consists of a series of samples selected on a sample basis. There is always the risk that sampling is not well representative of reality. Some urge critical situations and circumstances may be excluded. To eliminate this problem, we can increase the hypothesis space, using different training sets. In this way, each DT (or group of trees) is associated with a different training set, and DT truly independent of each other. This reduces the risk of misleading data in a training set. The more training sets we use, the lower the probability of machine failure. It is, in fact, unlikely that the same misleading data will be present in all training sets [1].

Neural networks are becoming a trend prevalent method for solving ML and artificial intelligence (AI) problems. The variety of neural network types and their application requires expert knowledge to choose the most appropriate approach. Traditional and newly emerging neural network approaches are backpropagation neural networks and their variants, self-organizing maps and associated methods, and a relatively new technique, deep learning (DL). The most important technical issues include overfitting and its prevention through regularization, ensemble and multitask modeling, model interpretation, and estimation of the applicability domain. Neural networks continue to grow in importance for biomedical applications. Recent developments in DL suggest further improvements may be gained in the analysis of large data sets. It's anticipated that neural networks will be more widely used in the future, and applied in non-traditional areas such as drug delivery systems, biologically compatible materials, and regenerative medicine [2]. Conventional neural networks utilize all the dimensions of the original input patterns for training and classification. However, a particular attribute of the input patterns does not necessarily contribute to classification and may even cause misclassification in some instances. A competitive EL method using the reduced input dimension, in contrast to the previous ensemble neural networks which adjust learning parameters, takes advantage of the information in each dimension of the input patterns. Since the degree of contribution of each attribute to classification is not known beforehand, the different input data sets with one dimension reduced are presented to multiple neural networks. The classification information from each competitive learning neural network is then combined to make a final decision for classification. To improve classification accuracy, the ambiguous output neurons are eliminated, which cannot be assigned to any class after training [3].

EL consists of generating a collection of classifiers whose predictions are then combined to yield a single unified decision. Ensembles of complementary classifiers provide accurate and robust predictions, which are often better than the predictions of the individual classifiers in the ensemble. Nevertheless, ensembles also have some drawbacks: typically, all classifiers are queried to compute the final ensemble prediction. Therefore, all the classifiers need to be accessible to address potential queries. This entails larger storage requirements and slower predictions than a single classifier. Ensemble pruning techniques that are useful to alleviate these drawbacks. Static pruning techniques reduce the ensemble size by selecting a sub-ensemble of classifiers from the original ensemble. In dynamic pruning, the querying process is halted when the partial ensemble prediction is sufficient to reach a stable final decision with a reasonable amount of confidence. The combination of static and dynamic pruning techniques provides a notable reduction in the memory requirements and an improvement in the classification time without a significant loss of prediction accuracy [4]. For years, we have witnessed the progressive introduction and adoption of AI in every sector of work, economics, and in the life of each individual.

In some cases, it is so integrated into our daily life that it is hard to notice its presence and influence. ML techniques have also made substantial progress in many sectors of industry and science, such as in Medicine, where it is even considered the great hope of the 21st century to improve the perspectives of life of humanity. In this field, it has already brought significant transformations in the health system. It is assumed that it will lead to further progress in the coming years: it will provide to physicians increasingly secure and efficient support in gathering, organizing, and analyze clinical data, making early diagnoses, and finding better solutions for the patients. There are predictions that ML will determine substantial changes of mind in healthcare, particularly in the disciplines that require more precise prognostic models, as in oncology, and in the ones based on pathological recognition, as in radiology. Although the introduction of the use of these algorithms has brought improvements and benefits in Medicine, there is still no certainty that their massive and generalized use could guarantee an increase in the efficiency of clinical activity. The result of the predictions of the algorithms can sometimes be misleading and controversial, leading the physician to wrong decisions.

HISTORICAL OVERVIEW

In general, in the history of mankind, the introduction of new technologies has never been simple and without unwanted effects. In the last century, man has tried to build machines that could help him in the calculation and his cognitive activities. In 1936, Alan Turing defined the concepts that still represent the hinges of current computers [5]. In 1943, McCulloch and Pitts created the first work that can be traced to AI, or a system that used artificial neurons with an "on" and "off" status, and a switch to "on" thanks to a certain number of stimuli generated by surrounding neurons [6]. Seven years later, Misky and Edmons built the first neural net named "Snark" [7]. In 1950, Turing proposed a test to see if a computer could be intelligent (Turing Test) through a specific experiment [8]. The first computer scientist in history to talk about ML was Arthur Lee Samuel, a pioneer in the field of AI. To demonstrate his ideas, he used a checkers program that measured the relationship between the computer's moves and what the player expected [9]. Only in 1956, John McCarthy coined the word "AI" [10]. Subsequently, the history of AI was characterized by ups and downs. In particular, significant steps have been taken about mathematical models (more and more sophisticated to imitate some brain functions such as pattern recognition, for example), but swinging from hardware and neural networks. In 1958, Frank Rosenblatt proposed the first neural network model, that is the so-called "Rosenblatt perceptron" [11].

The first significant turn from the technological point of view came between the end of the 70s and the decade of the 80s with the development of graphics processing units, which reduced the training times of the networks considerably, lowering them by 10-20 times. As we can see, today ML pervades many areas. For example, Google is using ML to create a cloud service that can offer personalized and non-personalized services. This project also includes Gmail, a famous email platform. All ML revolves around experience. Machines learn and perform a task better after they have performed it. The rule applied is: "you learn from your mistakes." Obviously, in the medical field, we try to minimize the margin of error. By inserting a few commands in an algorithm, machines develop their logic and operate at their best. We don't even imagine all the positive aspects that ML could have on our health. But some examples of ML have already become commonplace. Think of the Apple Watch, the smartwatches from the Apple company that measure various parameters of our health such as pressure, heart rate, and more. With this operation, a simple wrist watch can warn you that something is wrong with your body and prevent complications, as Apple herself shows in her promotional videos. Thanks to AI, it is then possible to keep the patient's medical history. A sort of collection of "big data" in the medical field. By crossing these data, machines may be able to derive crucial information for people's health. In Russia, the government developed software called "RadIO" for the early detection of lung cancer in collaboration with Github. The tool also acts at a higher level. This tool analyzes the data of a particular patient and also compares them with those of some people with lung cancer. In this way, the probability of preventing this disease is much higher than the norm [12]. The American startup Art Medical used AI to create an algorithm capable of monitoring the health of the hospitalized patient and the conditions of the instruments used to treat him (for example, drip or breathing ducts), alert doctors on any malfunctions [13].

STATE-OF-THE-ART

Related to the discipline of AI are two areas of study: ML and DL. ML, or automated learning, allows computers to "learn" and is, in fact, a way to achieve AI. We live in a world where technology imaginatively walks by our side and accompanies us in almost all the actions we carry out. Often, however, we find ourselves associating technology with something futile, which does not bring us many benefits. Going even further into the definition of the term "ML," we can say that it occurs when the machines manage to learn something without having been previously programmed. The term, in fact, was coined by Arthur Samuel in 1959, as "the ability to learn without being explicitly programmed" [14]. The central element of ML is the algorithm, which can be defined as a "procedure, implemented through an orderly and definite sequence of elementary steps, which leads to a well-determined result in a defined time." The system trains the algorithm in order to learn how to perform the activity, providing the algorithm with a large amount of data that allow it to correct or adapt, so that it can improve its performance more and more.

ML is characterized by some learning models and it is precisely on the basis of these models that it is possible to make a classification of the algorithms:

1) supervised learning;
2) unsupervised learning; and
3) reinforcement learning.

Supervised learning is an automatic learning technique, whose goal is to instruct the system so that it can solve tasks independently based on a set of previously assigned examples, formed by desired input and output pairs. Examples are provided to the computer in the form of possible inputs and their respective desired outputs, and the goal is to extract a general rule that associates the input with the correct output. Fundamental in supervised learning is the training set in which an example is presented: an answer (output) is automatically generated about the input and, if the answer matches to the exact one, then there is a reinforcement of the connections that led to the exact result. Otherwise, i.e., if the answer is not the expected one, the weights of the connections are modified in order to optimize the result and decrease the deviation from the correct solution. Then we move on to a test phase in which the generalization ability of the algorithm is verified (test set). The test set is used to determine the accuracy of the template. So, supervised learning is the process of data mining for deducing rules from training datasets. A broad array of supervised learning algorithms exists, every one of them with its advantages and drawbacks. Some basic issues affect the accuracy of classifier while solving a supervised learning problem, like bias-variance tradeoff, dimensionality of input space, and noise in the input data space. All these problems affect the accuracy of classifier and are the reason that there is no global optimal method for classification. There is not any generalized improvement method that can increase the accuracy of any classifier while addressing all the problems stated above [15].

The classification algorithm is used in multiple supervised learning methods and consists of identifying which category a new observation belongs to, based on the training set where the instances already belong to a category. So, the classification algorithms, starting from a training set, build a model that will later be used to classify new applications. A classification algorithm is known as "classifier." There are several classification algorithms, the main are the following:

1) *DT*: the goal of DT is to create a model that includes the value of a target variable based on different input variables. The tree is made

up of three parts:

a) internal node: it represents one of the input variables;
b) arc towards the child node: the possible value for that property; and
c) leaf: it represents the target value for the class, given the values of the input variables, represented by the root node to the leaf.

Subsequently, DT is used to determine to which class belongs the new instance you want to classify. With the growth of the depth and complexity of the decision-making rules of the tree will come a greater precision of the tree itself. DT is among the best classification models because:

a) it is not expensive to build;
b) it is simple to interpret and understand;
c) it can manage both numeric and category data; and
d) it has good accuracy in many applications, in comparison to other classification methods.

However, it has some disadvantages, as in the following examples:

a) in the case called "overfitting," by creating a tree too complex that does not generalize well with the training data; and
b) in the case of weak trees, where a small change in the data may involve a tree construction and different final forecasting;

ML approaches have broad applications in bioinformatics, and DT is one of the successful approaches applied in this field. DT are among the most popular of the new statistical learning methods being used in the pharmaceutical industry for predicting quantitative structure-activity relationships. Applications of DT in drug discovery research and extensions to the basic algorithm using hybrid or ensemble methods improve prediction accuracy [16]. By learning the algorithms of DT and ensemble classifiers, biologists can get the basic ideas of how ML algorithms work. On the other hand, by being exposed to the applications of DT and ensemble algorithms in bioinformatics, computer scientists can get better ideas of which bioinformatics topics they may work on in their future research directions [17]. Classification and regression trees, and their successors – bagging and random forests (RF) –, are statistical learning tools that are receiving increasing attention. However, due to the characteristics of censored data collection, standard classification and regression trees algorithms are not immediately transferable to the context of survival analysis. Questions about the occurrence and timing of events arise throughout psychological and behavioral sciences, especially in longitudinal studies. The prediction power and other key features of tree-based methods are promising in studies where an event occurrence is the outcome of interest [18]. Class imbalance presents a significant hurdle in the application of classification methods. A commonly taken approach is to learn ensembles of classifiers using rebalanced data. Examples include bootstrap averaging (bagging) combined with either under-sampling or oversampling of the minority class examples. However, rebalancing methods entail asymmetric changes to the examples of different classes, which in turn can introduce their own biases. Furthermore, these methods often require specifying the performance measure of interest *a-priori*, i.e., before learning. An alternative is to employ the threshold moving technique, which applies a threshold to the continuous output of a model, offering the possibility to adapt to a performance measure *a-posteriori*, i.e., a plug-in method. Surprisingly, little attention has been paid to this combination of a bagging ensemble and threshold-moving [19];

2) *pattern recognition*: K-nearest neighbors (KNN) is a useful method used for classifying objects based on the characteristics of the objects close to the one considered. It is considered the simplest of ML algorithms; and

3) *support vector machines (SVM)*: they are classification methods that generate a global approximation of the classification model using training data [20].

Unsupervised learning is a technique in which a series of inputs which must be reclassified according to standard features and then reasoning on subsequent inputs are provided to the system. Unlike supervised learning, data are not classified, but later the classes must be automatically discovered. Unsupervised learning techniques are very good with elements of a numeric type, but less efficient with non-numeric data. Some unsupervised learning algorithms are:

1) clustering: search for groups of objects so that objects belonging to a group are "similar" to each other and different from objects in other groups. In supervised learning, this was done by classification. The goal of clustering is to find the largest number of groups with homogeneous features, but which are as different as possible from each other. In detail, maximize the variance between clusters, but minimize it within. In biology, it can be used to derive taxonomies of animals and plants. In marketing, it can be used to derive and characterize groups of consumers. Clustering techniques can follow two philosophies:
   a) from bottom to top (aggregative or "bottom-up" methods): initially, all clusters are considered in their own right, subsequently the algorithm unites the closest clusters. The deadline is not achieved until we obtain an established number of clusters, or at least the minimum distance between clusters does not exceed a certain value or a pre-fixed value; and
   b) from top to bottom (divisive or "top-down" methods): as opposed to the bottom-up one, initially all the candidates are in a single cluster, and then the algorithm divides them into various clusters. The goal is to obtain increasingly homogeneous groups. The method stops when an established number of clusters has been reached.

   Clustering is a powerful and commonly used technique that organizes and elucidates the structure of biological data. Clustering plays a crucial role in several application domains, such as bioinformatics. In bioinformatics, clustering has been extensively used as an approach for detecting interesting patterns in genetic data. One application is population structure analysis, which aims to group individuals into subpopulations based on shared genetic variations, such as single nucleotide polymorphisms (SNP). Advances in deoxy-ribonucleic acid (DNA) sequencing technology have facilitated the obtainment of genetic datasets with exceptional sizes. Genetic data usually contain hundreds of thousands of genetic markers genotyped for thousands of individuals, making an efficient means for handling such data desirable [21]. Clustering data from gene expression, and metabolomics and proteomics experiments has proven to be useful at deriving a variety of insights, such as the shared regulation or function of biochemical components within networks. However, experimental measurements of biological processes are subject to substantial noise – stemming from both technical and biological variability – and most clustering algorithms are sensitive to this noise. Several methods of accounting for noise when analyzing biological data sets through clustering have been explored. Using a toy data set and two different case studies – gene expression and protein phosphorylation – the sensitivity of clustering algorithms to noise has been demonstrated. Several methods of accounting for this noise can be used to establish when clustering results can be trusted. These methods span a range of assumptions about the statistical properties of the noise and can therefore be applied to virtually any biological data source [22];

2) association rules: it is an unsupervised learning method whose purpose is to find exciting relationships among variables in large databases. This method is used on large volumes of data where there is the "transaction" concept. Association rules are a kind of "implications." The rule $X \rightarrow Y$ is interpreted as "in the transactions in which it appears X also appears Y." X is said "body," Y is said "head." In association rules we have two statistical measures:
   a) support: indicating the percentage of transactions that contain both X and Y; and

b) confidence: given the transactions that contain *X*, it indicates which is the percentage containing *Y*.

An example would be the supermarket receipt:

a) bread, butter → milk;
b) support 0.05; and
c) confidence 0.9.

It affirms that:

a) bread, butter and milk appear together in 5% of the receipts; and
b) the receipts that contain bread and butter together, 90%, will also contain milk.

Reinforcement learning is a ML technique that aims to achieve and implement algorithms in which its software agents are able to undertake actions in an environment in order to maximize their reward. The fundamental problem remains: what actions should I take to maximize my future reward? The answer is not simple because there are actions that are useful in the immediate but counterproductive in the future, on the other hand, actions that seem useless today could be beneficial in the future. Furthermore, the environment is usually stochastic, so we cannot be sure that the action we are going to perform will be productive. An important dilemma in this learning is that of "exploration vs exploitation," in the sense that we wonder how much we need to explore the environment looking for new (potentially better) strategies instead of concentrating on those learned. In many real situations, for each of our actions, feedback on the goodness of our action is provided to us by the environment. Learning in this case means generating many random answers to observe which have the highest degree of goodness. Reinforcement learning is used in two classes of environments:

1) static (or stochastic) environments, in which for each input-output pair there is a single reinforcement value; and
2) dynamic environments, where the timeline influences the interaction between agent and environment, modifies the value of the reinforcement, and alters the frequency of its availability to the

agent.

Two typical algorithms of reinforcement learning are:

1) temporal differences learning: it is an algorithm that changes with the variation of time to predict future rewards from current knowledges; and
2) Q-learning: it is a complementary method that does not require a model of the environment to predict the gain linked to each action; the environment is explored in a stochastic way and the reward is in the presence of the choice of action with immediate reinforcement and more excellent utility.

A novel heterogeneous ensemble selection approach based on the paradigm of reinforcement learning offers a more systematic and mathematically sound methodology for exploring the many possible combinations of base predictors that can be selected into an ensemble. The resultant ensembles are indeed substantially more parsimonious as compared to the full set of base predictors, yet still offer almost the same classification power, especially for larger datasets.

DL is an area of ML which is inspired by the structure and functioning of the biological brain, i.e., of the human mind. We can therefore consider ML as the model that "trains" AI, and DL as the algorithm that emulates the human mind. DL, in addition to algorithms, requires artificial neural networks (ANN); just these want to imitate the neural networks of the human brain, which are made up of networks of neurons that influence each other through synaptic connections. AI tries to imitate these principles to develop ANN. ANN have been used in various areas such as computer vision, automatic translation, voice recognition, medical diagnosis, and video games. Each neuron detects conditions and communicates them through discharge frequency. An artificial neuron, therefore, tries to emulate the characteristics of a neuron, which can be summarized in the following list:

1) neurons are strongly connected to each other;
2) each connection has a weight, which indicates the strength of the connection, and can have both a positive and a negative value;

3) the input that a neuron receives is given by the product between the signal coming from that neuron and the weight of the connection;
4) the total input of the neuron is the sum of the activations that the neuron receives from nearby neurons; and
5) the output of the neuron is sent to the "final" ones.

The architecture of an ANN is structured in a specific way:

1) units that receive inputs directly from the environment, called "the input layer";
2) those that produce the final output, called "the output layer"; and
3) the intermediate units, called "hidden units."

The way neurons are connected together can be different:

1) feed-forward networks, with unidirectional connections only from input to output;
2) recurring networks, with bidirectional connections; and
3) totally recurring networks, with also lateral intralayer connections.

During the initialization of a neural network, the value of the synaptic weights is randomly assigned, and then changed as the machine learns. The main objective, that is, learning, is that of modifying, action after action, the value of the weights, to make the next prediction more accurate than the previous one. The learning algorithms of ANN are:

1) *Hebb's rule*: "if two neurons connected are simultaneously active, the synaptic effectiveness of the connection is reinforced" [23]. If we associate an input *x* and an output *y*, the formula can be expressed as: $\Delta w = n\ (constant) * y * x$. Variants of Hebb's rule, but that however are all based on the same reinforcement principle according to discharge coordination:

a) postsynaptic rule: synaptic efficiency (the value of weight) decreases when the postsynaptic neuron is active, but the presynaptic one is not;
b) presynaptic rule: synaptic efficiency decreases when the presynaptic neuron is active, but the postsynaptic one is not; and
c) covariance rule: it is the same as Hebb's one, but it works with bipolar (-1/1) and non-binary (0/1) values; this leads to reinforcing the connection when the activity of the two neurons is correlated while it weakens it when it is unrelated;

2) *perceptrons*: the idea of correcting the weights of the network based on the error, calculated as the difference between the expected result and the obtained result, was elaborated by Frank Rosenblatt. Rosenblatt's perceptrons were neural networks with two layers of connections, in which the first one brought the information to the input layer, and the second was subjected to learning. For "perceptron" we indicate a network with a single unidirectional layer from input to output nodes. In a perceptron, the output activates when the weighted summation of the product between w and x is greater than zero, otherwise it will not activate. Then, there is the comparison between the value obtained from activation and the expected output; if these are different, there is a change in synaptic weight based on response and learning rate [11];

3) *the delta rule*: it is based on changing connections so as to reduce to zero the error value determined by the difference between the expected result and obtained result. This rule allows to describe performance with a function which measures the network error based on the median quadratic deviation between desired response and actual output; learning consists in decreasing the error value by varying the value of the synaptic connections;

4) *the function of the gradient's descent*: it allows to determine the minima and the maxima in a multi-variable function. In general, the

gradient of a function f, denoted by Δf, is defined at each point by the following relationship: for any vector $\vec{v}$, the scalar product $\vec{v} * \Delta f$ gives the value of the directional derivative of f with respect to $\vec{v}$. The research-based on the descent of the gradient determines the weight vectors that minimize E (network error). At each step, the weight vector is changed in the direction that produces the steepest descent of the gradient on the surface of the error; and

5) *back-propagation of the error*: it is drafting of the delta rule which allows to train multilayer networks. The algorithm allows us to modify the weights of the connections in such a way that a certain error function E is minimized. The algorithm compares the system output value with the desired value (target). On the basis of the calculated difference (error), the algorithm changes the synaptic weights of the neural network, making the set of the output values gradually converge towards those desired.

Although the use of ensemble methods in ML is ubiquitous due to their proven ability to outperform their constituent algorithms, ensembles of optimization algorithms have received relatively little attention. Existing approaches lag behind ML in both theory and practice, with no principled design guidelines available. Randomly-composed ensembles can outperform ensembles of high-performing algorithms under certain conditions, and judicious choice of diversity metric is required to construct good ensembles [24].

As we have seen, AI is a field of computer science that deals with making machines that can solve problems independently, without the help of man. It is demonstrated when an activity, first carried out by a human being who has the ability to learn and reason, may be performed independently by a machine or more generally by an intelligent system. Starting from the functioning of the human brain, of which we still do not fully understand the mechanisms, an AI system must guarantee certain actions/functions:

1) acting humanly (i.e., indistinctly with respect to a human being);
2) thinking humanly (i.e., solving a problem with cognitive functions);
3) thinking rationally (i.e., using logic like a human being does); and
4) acting rationally (i.e., by starting a process to obtain the best result expected on the basis of the available information).

There are two different currents that have created a sort of debate over the years, to define the concept and the operation of an intelligent machine:

1) *weak AI*: technological systems capable of simulating human skills, without completely reaching the typical human intellectuality, for example the vocal assistants; and
2) *strong AI*: we speak of "wise systems" that can develop their own intelligence, not simulating that of man but developing one independently; these systems are defined "aware of themselves."

This debate was brought to a more concrete level when in 1950 Alan Touring published an article entitled "Computing Machinery and Intelligence," within which he gave the definition of intelligence of a machine based on a test called "The Imitation Game" or even "Touring test." The test consists of a game involving three people: one man (*A*), one woman (*B*), and a third individual (*C*). *C* is kept separate and has to guess, asking a series of questions, the sex of *A* and *B*; on the other side, *A* will have to try to deceive *C*, instead, *B* will have to help him. The assumption of the test is that machine replaces *A*; if the number of correct answers is equal even after the replacement, then the machine will be considered intelligent since it would be indistinguishable from the human being. The conclusion reached by Touring was the definition of intelligent machine or a machine capable of thinking, capable of elaborating ideas, of relating them to others, and of knowing how to express them.

The functioning of an AI system, from an intellectual point of view, has four different functional levels:

1) *understanding*: through the simulation of cognitive skills and being able to recognize images, videos, texts, and extrapolate information;
2) *reasoning*: thanks to logic, the systems manage to connect the multiple information collected, using specific algorithms;
3) *learning*: through specific functions for reading inputs, systems can give a correct output (e.g., ML); and
4) *interaction*: mode of operation of AI about its interaction with humans, with a strong growth of natural language processing systems, technologies that allow humans to converse with machines using natural language.

AI is now an integral part of our society and has found its use in different industrial sectors, using at the various level the previously expressed functions:

1) *finance*: for years, trading algorithms capable of make autonomous decisions, thanks to the ability to process large amount of data in much less time than a human being, have been introduced. In addition, applications such as the Digit app, which help and support the consumers to optimize their expenses based on their own personal habits and goals, have been presented;
2) *automation and industry*: for tens of years, robots have been replacing human beings in repetitive and dangerous jobs, but the true evolution of AI in this sector is building machines autonomously driven with systems capable of detecting, identifying, and processing events and information in real time that allow replacement of human guidance;
3) *daily life*: applications such as Siri, Cortana, Alexa, etc... assist every day each individual in the search for information through an intelligent interaction with the human being. Also in the video ludic field, AI algorithms are widely used to maximize the realistic effect of the game and the "human" behavior of the different characters;

4) *defense*: in recent years, robots and drones capable to undertake autonomous activities to replace the soldier in military actions have been introduced;
5) *advertising*: advertising messages are increasingly targeted to the individual thanks to specific algorithms capable of processing in real-time consumer's habits and preferences;
6) *medicine*: this sector represents one of the major areas of interest and diffusion of AI in support of the activity of the clinical staff. Many hospitals and medical centers have equipped themselves to make Medicine and technology interact.

MEDICAL APPLICATIONS

Introduction

Progress in biomedical research has resulted in an explosive growth of data ("big data"). Use of the world wide web (WWW) for sharing data has opened up possibilities for exhaustive data mining analysis. Symbolic ML approaches used in data mining, especially ensemble approaches, produce large sets of patterns that need to be evaluated. Manual evaluation of all patterns by a human expert is almost impossible. A new approach to the evaluation of ML-induced knowledge by introducing a pre-evaluation step has been proposed. Pre-evaluation is the automatic evaluation of patterns obtained from the data mining phase, using text mining techniques and sentiment analysis. It is used as a filter for patterns according to the support found in online resources, such as publicly-available repositories of scientific papers and reports related to the problem. The domain expert can then more easily distinguish between patterns or rules that are potential candidates for new knowledge [25]. Big data is now on everyone's lips, even in healthcare. However, it is not always clear how the scientific data collected can actually be used in the medical and health sector. The first step is to understand what information can actually be obtained, thanks to technology, regarding citizens' preferences and concerns, and then to be able to use it to offer an even more effective health experience and promote research. This is the challenge of using big data in Medicine, even closer to a promise than to reality. According to the official definition of the European Community (EC), "big data in healthcare refers to large sets of data collected periodically or automatically, which are archived electronically, and reusable for the purpose of improving the performance of the healthcare system." It is therefore a fundamental aspect of any research activity, which has been transformed thanks to technological and informatics development. This is where ML algorithms intervene, where a researcher needs technological support to extract meaning from the data. There are thousands of terabytes of medical and scientific data, distributed by research projects, publications, and digital medical records. If analyzed carefully and with the appropriate tools, these can give the right boost to research in the health sector, not only as regards the production of new drugs, but also for the choice of a new clinical treatment. Therefore, a potentially endless database is created, enriched every day through e-health services, such as the electronic health record (EHR) of each patient, which are inserted and archived thanks also to the medical management software used by healthcare facilities and health professionals. Among the many information that can be collected about patients, there are also indications of preferences that may seem little close to the health sector, but which reveal interesting information taken directly from online behaviors. A Google search, the content of an email, and shared posts on social media can tell citizens' concerns and priorities. The most classic example is that of Derrick de Kerckhove, who, in 2011, thanks to the use of big data, discovered that people feared the outbreak of a new epidemic rather than terrorism [26].

Once big data is made available to the physician, pharmacist, or healthcare professional, the problem arises of how to use it, which in turn hides a crucial theme: that of data classification. On the one hand, there are personal data, therefore relating to the patient-citizen, and on the other hand, organizational data, which describe the activity of the institution responsible for protecting the health of the citizen, whether it is a local health unit or a clinic. The combination and a prudent and competent use of this information can help the healthcare professional in various individual or collective activities, summarized in six points:

1) complexity and precision medicine;
2) taking care of patients, in particular those suffering from chronic diseases, or those requiring continuity of care over the medium and long term;
3) socio-health care continuity;
4) predictive and preventive medicine;
5) overall reorganization of health processes from the local plan up to the national health service (NHS); and
6) open innovation, to foster dialogue and shared planning between those involved in health care and those who, on the other hand, dedicate themselves to state and corporate welfare.

The potential of big data in Medicine is enormous and those who have been studying the sector for years are convinced that it can actually profoundly innovate and improve healthcare. The positive fallout will, of course, be on the patient. Knowing more information on his state of health, in fact, would allow everyone to be followed in an increasingly personal and "tailor-made" way. For example, thanks to big data it is possible to collect enough data to allow physicians to know 12 months in advance and with a 98% safety percentage if a certain drug will cause side effects to a certain patient. Information that can be used to prescribe a different drug, helping the patient to heal earlier, but also to save money. To make another example, careful use of big data also allows the citizen to be made responsible for his health because people can authorize the sharing of some daily information such as the rhythm of sleep or the heartbeat. Any anomaly is, therefore, immediately detected and treated.

Digital Healthcare

The digitization process in Italy is slow in healthcare as in all other sectors. Tools such as the EHR and the new healthcare call center services are spreading, albeit not in a completely homogeneous way. All innovations that go in the direction of the dematerialization of information and therefore, also of the collection of integrated information available to physicians. However, the effective integration of this information with web analytics and with other information about the patient-citizen that can actually already be collected is still lacking. There are pilot projects that are active in this sense, but their activity remains residual and still not widespread throughout the national territory. It is therefore, necessary to have an integrated approach that takes into consideration a wider data collection spectrum, but also greater attention to the professionalism necessary to manage and coordinate big data. They are called "data scientist" and "data analyst" and are new professions, which combine mathematical, technological, and statistical skills made available to an organization to allow it to grow and improve. At the same time, they also contribute to the development of platforms and software for managing and analyzing big data in Medicine. And where the amount of data to be analyzed exceeds the researcher's skills, ML then comes to the aid of the sector by offering tools that allow the best interpretation of the meaning of the data, thus guaranteeing a richer future of information for medical and health research [27].

Digital healthcare is increasingly involved in the use of DL AI systems. In the area of imaging diagnostics in particular, studies are focusing on DL architectures capable of processing huge amounts of data. However, these are processes on which questions related to the "unknowability" of the algorithm can weigh: this is where the new concept of "explainable AI" comes from. This is an increasingly "necessary" concept, given that information and communication technology products attributable to AI are now widespread, even if there is no full awareness of this fact by the public of non-experts. Even a simple keyword search on a search engine implies the use of products of this type. The same applies to the organization of the personal page in the most popular social networks, based on the activities of the page's owner, in order to make the browsing experience more rewarding. In this context, it is not surprising that there is a growing interest in AI applications in Medicine: in particular, the field of diagnostic imaging (especially radiology, but also histology and histopathology) is suitable to the analysis of large quantities of data, typical of ML (or of its most recent version, DL).

Diagnostic Imaging

The potential of AI in Medicine is however very consistent also in other areas, such as the automatic analysis of medical records which, being compiled in an unstructured way, have traditionally not been considered as exploitable with an algorithmic approach, up to automatic and structured composition of reports; the progress of AI are instead allowing to use these data too, no longer so "difficult" for the software as it was up to the recent past. Regarding diagnostic imaging, the analysis algorithms can reveal features that even an expert eye is not able to evaluate: almost invaluable differences in intensity (as well as pixel patterns, or textures) can be reliably quantified by software tools, which have the additional advantage of offering constant performances. Another certainly feasible goal for AI in Medicine will be the integration of data from various sources: if diagnostic imaging ("radiomics" derived from radiology, nuclear medicine, nuclear magnetic resonance, ultrasound, etc.) gives indications on the "phenotype," that is the set of morphological and functional characteristics of an organism determined by the interaction between its genetic constitution and the environment, the analysis techniques of the "genotype," i.e., the genetic constitution, the hereditary heritage of an individual, provide other valuable information (e.g., through genetic sequencing of an individual or a group of population), which can be combined with that relating to the phenotype (e.g., "radio-genomics"). We are therefore witnessing the exponential growth of health data (and also the simple well-being of the person, see the diffusion of wearable fitness devices), so the traditional analysis techniques are not suitable for satisfactorily managing this mass of data. AI applications (especially DL), on the other hand, are naturally predisposed to cope with this explosion of data, as they always work better as the amount of training data increases, a necessary phase to build the optimal neural network for a given clinical problem.

As already mentioned, the most common AI paradigm is represented by ML techniques. Among these, the most popular architecture currently is given by DL, in which a neural network of tens or even hundreds of layers of "neurons," or elementary processing units, is used. The complexity of the DL architectures means that they behave like "black boxes," so it is practically impossible to identify the exact mechanism by which the system provides certain answers: the system's response to the stresses received during training translates in fact in the continuous refinement of a large amount of parameters (related to the interconnections between the "neurons" of the network), and it is not possible – except with ultra-specialistic analyzes – to predict the "macroscopic" behavior of the network (e.g., in terms of clinical decision) on the basis of the value of these parameters which constitute their "microscopic" substrate. Given the widespread use of DL solutions, this problem will become increasingly felt. It should be stressed that in the medical field the accountability, or responsibility, of the professional is of primary importance: any medical decision must be justified in the aftermath, possibly through objective evidence. The same must apply when the outcome of an AI processing also contributes to the clinical decision, so that "black box" architectures are hardly compatible with the health sector; furthermore, since these software applications must be certified, the criticality of this procedure can be understood in the face of an algorithm that cannot be simply explained. The need to have transparent, understandable, and explainable clinical decisions also joins the requirements of the new European General Data Protection Regulation (GDPR), which restricts the use of "opaque" data processing: even without the obligation to explain every single step, however, there is the need to be able to trace the elaborations made, if it becomes necessary in the future. The concept of "explainable AI" is thus affirming itself: it is important to be able to identify the path taken by an AI system to arrive at certain conclusions and to be sure that the performed "reasoning" is acceptable. Specifically for health applications, the use of "black boxes" certainly poses non-trivial problems, therefore, in defining certification, quality assurance, and standardization paths for potential medical devices based on DL [28]. The use of ensembles in ML has had a considerable impact in increasing the accuracy and stability of predictors. This increase in accuracy has come at the cost of comprehensibility as, by definition, an ensemble model is considerably more complex than its component models. This is of significance for decision support systems in Medicine because of the reluctance to use models that are essentially black boxes. Work on making ensembles comprehensible has so far focused on global models that mirror the behavior of the ensemble as closely as possible. With such global models, there is a clear tradeoff between comprehensibility and fidelity. Looking at local comprehensibility, where the output of the ensemble is explained on a case-by-case basis, this meets the requirements of medical decision support systems. This approach identifies the ensemble members that best fit the case in question and presents the behavior of these in explanation [29].

The future of biomedicine will be based on the combination of different disciplines and big data. Past experience and the data of millions of patients can help physicians make faster and more accurate diagnoses, facilitate researchers to understand the genetic mechanisms underlying diseases, and predict the risk of a disease in time useful to prevent it. New technologies are transforming Medicine and this revolution starts from data. Health data, clinical images, genome sequences, data on prescribed therapies, and results obtained: data that each of us has helped to create. Although the first signs of AI in Medicine date back to the 1980s, it is only with the beginning of the new millennium that there has been an explosion of interest in this sector worldwide. Fifty years ago, expectations were very high: these technologies should have almost replaced most of the physician's functions, but in 2020 we can hardly deal effectively with EHR. ML is one of the main branches of AI and is a fundamental technology for the management and understanding of the huge amount of data, health and otherwise, that we produce daily. In general, ML includes different mechanisms that allow an intelligent machine to learn and improve its skills over time thanks to experience. Everything is based on algorithms. An algorithm is a procedure that allows the resolution of a problem by applying specific instructions previously provided. In ML, a model learns from the examples available – input and output data are provided for each task so that it learns the link between them – and it does not simply follow the pre-established rules with which it has been programmed, but generalizes the process to apply it to data never seen before. An easy-to-understand example in healthcare is that of images, including magnetic resonance imaging (MRI), X-rays, ultrasound scans, and more. Medical images are among the most important tools that doctors use to diagnose anomalies and pathologies, ranging from simple bone fractures to cancer. However, analyzing medical images can be a difficult and time-consuming process. For example, if we provide our ML algorithm with the initial data, in this case the images, and combine them with information indicating a suspected lesion, the model will be able to find repeated patterns (i.e., associate the same response with a given type of image) and to provide the correct answer. Often more likely than the physician. It is no coincidence that one of the first authorizations of the United States (US) regulatory agency – Food and Drug Administration (FDA) – for the use of AI in Medicine concerns the analysis of images. Specifically, it is an algorithm capable of analyzing photographs of the retina for the identification of diabetic retinopathy. The computational approach can be applied not only to image detection techniques, but also for the understanding of genetic data, which cannot be interpreted with the naked eye, or for drug development, reducing the time and costs of preliminary investigations thanks to simulations in silico (i.e., computer with mathematical models) of the molecule's activities. The data relating to diagnoses and therapies are available in the form of medical records, and all that needs to be done is to enter the data combined with the correct diagnosis, so the algorithm can learn. Obviously, this is an oversimplification, but the fact remains that it can be useful to improve the time needed for diagnosis, its accuracy (in the database there are more cases to compare than those comparable with human skills only) or to facilitate the identification of specific pathologies also by non-specialized physicians, so as to subject the patient to specialist examinations as soon as possible in case of need. Also, each patient suffers a diagnosis error at least once in his life and this delays or complicates treatment. It is crucial to emphasize that this does not only happen for rare diseases, but also for the most common cases of heart problems or simple infections. Thanks to the processing of the data collected during the visits, it will be possible to identify the problem earlier and better respond to the patient's needs. Also applicable to surgery, ML techniques can be adapted for real-time analysis of video footage taken during surgical operations, so as to make surgical practice safer, for example by highlighting problems that the human eye could escape or reducing cases of human error.

Genomics

Speaking of data, those on genomics are accumulating in unprecedented quantities, giving scientists the opportunity to study genetic variants, related to specific diseases or not. As the cost of DNA sequences becomes increasingly accessible, the study of our genome seems to be one of the main ways of directing us towards personalized medicine, but the analysis of these data is not obvious at all. And having so much data without the possibility of processing it is equivalent to not having it at all. ML, as well as DL – that is, a class of ML algorithms that use ANN capable of learning extremely complicated relationships between inputs and outputs provided and which exceed human skills in performing tasks such as image classification – will be fundamental for this revolution. In coming years, medical practice will be upset by the exponential growth of genetic, molecular, biometric, and chemical data collected using sensors and other medical devices during exams and hospitalizations, but also thanks to the increasingly widespread use of the so-called "wearable devices," like the well-known health watches that measure heart rate, rhythm and hours of sleep, pressure, or other vital parameters. However, the limits, mostly linked to data acquisition, should not be underestimated. The still limited use of EHR, often due to the difficulty of use, limits the collection and sharing of digital data. Another problem is that the data on which the learning of the algorithm is based derive from standard medical practice, which however does not always correspond to the ideal solution for the patient. Furthermore, the data is not collected uniformly for all sections of the population, as in the case of the differences in treatment that American patients have based on the possibility or not of having health insurance. ML can help diagnose diseases and other health problems. AI leads to the creation of clinical decision support system or systems that support physicians in their clinical decisions. Listening to the opinion of more experienced physicians, this is a practice that clinicians often follow. Confronting those who have worked in healthcare for a long time, a physician can choose the best cure for a patient. This is precisely the basic idea of the Health Cloud software, created by GE Healthcare. A cloud that connects its diagnostic imaging devices and allows physicians who use it to exchange information and opinions [30]. ML, and in particular DL, is revolutionizing the approach to diagnosis [31]. It allows to match and, in some cases, exceed the physician's performance (not to be mistaken with the "to be able to replace it"). For example, in the early diagnosis of diabetic retinopathy on images of the ocular fundus [32], in the diagnosis of melanomas [33], in the detection of brain lesions on MRI [34], in the diagnosis of pneumonia on X-ray images [35], in prenatal monitoring on ultrasound images [36], and others that are currently being studied.

Genome-wide association studies using thousands to hundreds of thousands of SNP markers and region-wide association studies using a dense panel of SNP are already in use to identify disease susceptibility genes and to predict disease risk in individuals. Because these tasks become increasingly important, three different data sets have been provided for the Genetic Analysis Workshop 15 (GAW15), thus allowing examination of various novel and existing data mining methods for both classification and identification of disease susceptibility genes, gene by gene or gene by environment interaction. The approach most often applied in this presentation group has been RF because of its simplicity, elegance, and robustness. It was used for prediction and for screening for interesting SNP in a first step. The logistic tree with an unbiased selection approach appeared to be an interesting alternative to efficiently select interesting SNP. ML, specifically ensemble methods, might be useful as pre-screening tools for large-scale association studies because they can be less prone to overfitting, can be less computer processor time intensive, can easily include pair-wise and higher-order interactions compared with standard statistical approaches, and can also have a high capability for classification. However, improved implementations that are able to deal with hundreds of thousands of SNP at a time are required [37]. RF is an ensemble classification algorithm which performs well when most predictive variables are noisy and can be used when the number of variables is much larger than the number of observations. The use of bootstrap samples and restricted subsets of attributes makes it more powerful than simple ensembles of trees. The main advantage of a RF classifier is its explanatory power: it measures variable importance or impact of each factor on a predicted class label. These characteristics make the algorithm ideal for microarray data. It was shown to build models with high accuracy when tested on high-dimensional microarray datasets. Current implementations of RF in the ML and statistics community, however, limit its usability for mining over large datasets, as they require that the entire dataset remains permanently in memory [38]. RF, however, has emerged as an efficient algorithm capable of handling high-dimensional data. RF provides a proximity measure that can capture different levels of co-occurring relationships between variables. RF has been widely considered a supervised learning method, although it can be converted into an unsupervised learning method. Therefore, RF-derived proximity measure, combined with a clustering technique, may be well suited for determining the underlying structure of unlabeled data. Applying a cluster ensemble approach, combining multiple RF clustering, produces more robust and higher-quality results as a consequence of feeding the ensemble with diverse views of high-dimensional genetic data obtained through bagging and random subspace, the two key features of the RF algorithm. Multivariate quantitative traits arise naturally in recent neuroimaging genetics studies, in which both structural and functional variability of the human brain is measured non-invasively through techniques such as MRI. There is growing interest in detecting genetic variants associated with such multivariate traits, especially in genome-wide studies. RF classifiers, which are ensembles of DT, are amongst the best performing ML algorithms and have been successfully employed for the prioritization of genetic variants in case-control studies. RF can also be applied to produce gene rankings in association studies with multivariate quantitative traits, and to estimate genetic similarities measures that are predictive of the trait. However, in studies involving hundreds of thousands of SNP and high-dimensional traits, a very large ensemble of trees must be inferred from the data in order to obtain reliable rankings, which makes the application of these algorithms computationally prohibitive [39]. Ensemble predictors such as the RF are known to have superior accuracy, but their black box predictions are difficult to interpret. In contrast, a generalized linear model (GLM) is very interpretable, especially when forward feature selection is used to construct the model. However, forward feature selection tends to overfit the data and leads to low predictive accuracy. Therefore, it remains an important research goal to combine the advantages of ensemble predictors (high accuracy) with the advantages of forwarding regression modeling (interpretability). To address this goal, several articles have explored GLM-based ensemble predictors. Since limited evaluations suggested that these ensemble predictors were less accurate than alternative predictors, they have found little attention in the literature. Comprehensive evaluations involving hundreds of genomic data sets, the University of California-Irvine (UCI) ML benchmark data, and simulations have been used to give GLM-based ensemble predictors a new and careful look. A novel bootstrap aggregated (bagged) GLM predictor that incorporates several elements of randomness and instability (random subspace method, optional interaction terms, and forward variable selection) often outperforms a host of alternative prediction methods, including RF and penalized regression models (ridge regression, elastic net, and lasso). This random GLM (RGLM) predictor provides variable importance measures that can be used to define a "thinned" ensemble predictor (involving few features) that retains excellent predictive accuracy. RGLM is a state-of-the-art predictor that shares the advantages of a RF (excellent predictive accuracy, feature importance measures, and out-of-bag estimates of accuracy) with those of a forward selected GLM (interpretability) [40]. Whole genome ribonucleic acid (RNA) expression studies permit systematic approaches to understand the correlation between gene expression profiles to disease states or different developmental stages of a cell. Microarray analysis provides quantitative information about the complete transcription profile of cells that facilitate drug and therapeutics development, disease diagnosis, and understanding in the basic cell biology. One of the challenges in microarray analysis, especially in cancerous gene expression profiles, is to identify genes or groups of genes that are highly expressed in tumor cells but not in normal cells and *vice versa.* Previously, it has been shown that ensemble ML consistently performs well in classifying biological data. Three different supervised ML techniques in cancer classification, namely C4.5 DT, and bagged and boosted DT, have performed classification tasks on publicly available cancerous microarray data and the classification/prediction performance of these methods has been compared. It has been observed that EL (bagged and boosted DT) often performs better than single DT in this classification task [41].

Microarray technology is increasingly being applied in biological and medical research to address a wide range of problems such as tumor classification. A crucial statistical question associated with tumor classification is the identification of new tumor classes using gene expression profiles. Essential aspects of this clustering problem include identifying accurate partitions of the tumor samples into clusters and assessing the confidence of cluster assignments for individual samples. Two resampling methods, inspired from bagging in prediction, have been proposed to improve and assess the accuracy of a given clustering procedure. In these ensemble methods, a partitioning clustering procedure is applied to bootstrap learning sets and the resulting multiple partitions are combined by voting or the creation of a new dissimilarity matrix. As in prediction, the motivation behind bagging is to reduce variability in the partitioning results via averaging. The performances of the new and existing methods have been compared using simulated data and gene expression data from two published cancer microarray studies. The bagged clustering procedures were, in general, at least as accurate and often substantially more accurate than a single application of the partitioning clustering procedure. A valuable by-product of bagged clustering are the cluster votes, which can be used to assess the confidence of cluster assignments for individual observations [42]. Stratification of patients for multidrug response is a promising strategy for cancer treatment. Genome-based prediction models have great potential for this purpose because the extent of drug sensitivity may be attributed to the heterogeneity of the underlying genetic characteristics of cancer.

However, microarray data is difficult to analyze and is not reproducible. Several ML algorithms have therefore, been developed in a repeatable manner. RF algorithms, which use an ensemble approach based on classification and regression trees, appear to be superior for predicting multidrug sensitivity. This is because ensemble methods are more effective when there are much more predictors than samples [43]. Determining the functional role of enzymes is very important to build the metabolic blueprint of an organism and to identify the potential roles enzymes may play in metabolic and disease pathways. With exponential growth in gene and protein sequence data, it is not feasible to experimentally characterize the functions of all enzymes. Alternatively, computational methods can be used to annotate the enormous amount of unannotated enzyme sequences. For function prediction and classification of enzymes, features based on amino acid composition, sequence and structural properties, domain composition, and specific peptide information have been widely used by different computational approaches. Each feature space has its own merits and limitations on the overall prediction accuracy. Prediction accuracy improves when ML methods are used to classify enzymes. Given the incomplete and unbalanced nature of annotations in biological databases, ensemble methods or methods that bank on a combination of orthogonal feature are more desirable for achieving higher accuracy and coverage in enzyme classification [44]. Many automatic classifiers were introduced to aid inference of phenotypical effects of uncategorized nonsynonymous single nucleotide variations (nsSNV) in theoretical and medical applications. Lately, several meta-estimators have been proposed that combine different predictors, such as polymorphism phenotyping (PolyPhen) and scale-invariant feature transform (SIFT), to integrate more information in a single score. Although many advances have been made in feature design and ML algorithms used, the shortage of high-quality reference data, along with the bias towards intensively studied *in vitro* models, call for improved generalization ability to further increase classification accuracy and handle records with insufficient data. Since a meta-estimator basically combines different scoring systems with highly complicated nonlinear relationships, DL (supervised and unsupervised), which is particularly efficient at discovering hierarchies of features, can improve classification performance. While it is believed that one should only use DL for high-dimensional input spaces and other models (logistic regression, SVM, Bayesian classifiers, etc.) for simpler inputs, the ability of neural networks to discover intricate structure in highly heterogeneous datasets can aid a meta-estimator. Even relatively simple modern neural networks can significantly improve both prediction accuracy and coverage [45].

Computer-Aided Diagnostics

Respiratory diseases account for a significant proportion of deaths and disabilities across the world. Chest X-ray analysis remains a common diagnostic imaging modality for confirming intra-thoracic cardiopulmonary abnormalities. However, there remains an acute shortage of expert radiologists, particularly in under-resourced settings, resulting in severe interpretation delays. Chest X-ray analysis is also a common part of the protocol for confirming active pulmonary tuberculosis (TBC). However, many TBC endemic regions are severely resource-constrained in radiological services impairing timely detection and treatment. These issues can be mitigated by a computer-aided diagnostic (CAD) system to supplement decision-making and improve throughput while preserving and possibly improving the standard-of-care. Systems reported in the literature or popular media use handcrafted features and/or data-driven algorithms like DL to learn underlying data distributions. The remarkable success of convolutional neural networks toward image recognition tasks has made them a promising choice for automated medical image analyses. However, convolutional neural networks suffer from high variance and may overfit due to their sensitivity to training data fluctuations. EL helps to reduce this variance by combining predictions of multiple learning algorithms to construct complex, non-linear functions and improve robustness and generalization. The performance of an ensemble of ML models applied to the challenge of classifying normal and abnormal chest X-ray can significantly reduce the diagnostic load of radiologists and primary-care physicians [46]. CAD tools can supplement decision-making while simultaneously addressing the gap in expert radiological interpretation during mobile field screening. These tools use hand-engineered and/or convolutional neural networks computed image features. Convolutional neural networks, a class of DL models, have gained research prominence in visual recognition. It has been shown that EL has an inherent advantage of constructing non-linear decision-making functions and improve visual recognition [47].

Confocal laser endomicroscopy is an advanced optical fluorescence imaging technology that has the potential to increase intraoperative precision, extend resection, and tailor surgery for malignant invasive brain tumors because of its subcellular dimension resolution. Despite its promising diagnostic potential, interpreting the gray tone fluorescence images can be difficult for untrained users. Confocal laser endomicroscopy images can be distorted by motion artifacts, fluorescence signals out of detector dynamic range, or may be obscured by red blood cells, and thus interpreted as nondiagnostic. However, just a single confocal laser endomicroscopy image with a detectable pathognomonic histological tissue signature can suffice for intraoperative diagnosis. Dealing with the abundance of images from confocal laser endomicroscopy is not unlike sifting through a myriad of genes, proteins, or other structural or metabolic markers to find something of commonality or uniqueness in cancer that might indicate a potential treatment scheme or target. The bioinformatical analysis methodology of confocal laser endomicroscopy images begins to assist the neurosurgeon and pathologist to rapidly connect on-the-fly intraoperative imaging, pathology, and surgical observation into a conclusion system within the concept of theranostics. Two major approaches include the models that can automatically classify confocal laser endomicroscopy images into diagnostic/nondiagnostic, glioma/nonglioma, tumor/injury/normal categories, and models that can localize histological features on the confocal laser endomicroscopy images using weakly supervised methods. Significant advances in speed and precision of automated diagnostic frame selection would augment the diagnostic potential of confocal laser endomicroscopy, improve operative workflow, and integration into brain tumor surgery. Such technology and bioinformatics analytics lend themselves to improved precision, personalization, and theranostics in brain tumor treatment [48]. In last years, several approaches to develop an effective CAD system for Parkinson's disease (PD) and Alzheimer's disease (AD) have been proposed. Most of these methods have focused almost exclusively on brain images through the use of ML algorithms suitable to characterize structural or functional patterns. Those patterns provide enough information about the status and/or the progression at intermediate and advanced stages of PD.

Nevertheless, this information could be insufficient at early stages of the pathology. The Parkinson's Progression Markers Initiative (PPMI) database includes neurological images along with multiple biomedical tests. This information opens up the possibility of comparing different biomarker classification results. As data come from heterogeneous sources, it is expected that we could include some of these biomarkers to obtain new information about the pathology [49]. AD is an incurable, progressive neurological brain disorder. Earlier detection of AD can help with proper treatment and prevent brain tissue damage. Several statistical and ML models have been exploited by researchers for AD diagnosis. Analyzing MRI is a common practice for AD diagnosis in clinical research. Detection of AD is exacting due to the similarity in AD MRI data and standard healthy MRI data of older people. Recently, advanced DL techniques have successfully demonstrated human-level performance in numerous fields, including medical image analysis. A deep convolutional neural network for AD diagnosis using brain MRI data analysis has been proposed. While most of the existing approaches perform binary classification, this model can identify different stages of AD and obtains superior performance for early-stage diagnosis [50].

Bioinformatics

Bioinformatics applies data mining, i.e., modern computer-based statistics, to biomedical data. It leverages on ML approaches, such as ANN,

DT, and clustering algorithms, and is ideally suited for handling huge data amounts [51]. Advanced data mining techniques such as DT have been successfully used to predict a variety of outcomes in complex medical environments. Furthermore, previous research has shown that combining the results of a set of individually trained trees into an ensemble-based classifier can improve overall classification accuracy. While applying an ensemble-based approach rather than a single classifier has a small positive effect on the accuracy rate, this also varies depending on the subject under analysis. Finally, scaling data before prediction also has a small positive effect on the accuracy rate for this dataset. A graphical analysis of the distribution of accuracy rates across ensembles supports these findings. Laboratories looking to include ML as part of their decision support processes need to be aware that the disease outcome, the ML method used, and the subject interact to affect the enhanced laboratory diagnosis of the disease, as determined by primary immunoassay data in concert with multiple routine pathology laboratory variables. This awareness will lead to the informed use of existing ML methods, thus improving the quality of laboratory diagnosis via informatics analyses [52]. Mass spectrometry has been used in biochemical research for a long time. However, its potential of discovering proteomic biomarkers using protein mass spectra aroused tremendous interest in last few years. In spite of its potential of biomarker discovery, it is recognized that identification of meaningful proteomic features from mass spectra needs careful evaluation. Hence, extracting meaningful features)and discriminating the samples based on these features is still an open area of research. Several research groups are actively involved in making the process as perfect as possible. No single classification algorithm tends to work well on all data. Also, the performance depends on the performance criteria used to judge several classifiers. Adaptive ensemble classifier which is constructed combining several good classifiers and optimized against an array of performance measures tends to have better predictive performance on the test samples [53]. Multiple computational methods for predicting drug-target interactions have been developed to facilitate the drug discovery process. These methods use available data on known drug-target interactions to train classifiers with the purpose of predicting new undiscovered interactions. However, a key challenge regarding this data that has not yet been addressed by these methods, namely class imbalance, is potentially degrading the prediction performance. Class imbalance can be divided into two sub-problems. Firstly, the number of known interacting drug-target pairs is much smaller than that of non-interacting drug-target pairs. This imbalance ratio between interacting and non-interacting drug-target pairs is referred to as the between-class imbalance. Between-class imbalance degrades prediction performance due to the bias in prediction results towards the majority class (i.e., the non-interacting pairs), leading to more prediction errors in the minority class (i.e., the interacting pairs). Secondly, there are multiple types of drug-target interactions in the data with some types having relatively fewer members (or are less represented) than others. This variation in representation of the different interaction types leads to another kind of imbalance referred to as the within-class imbalance. In within-class imbalance, prediction results are biased towards the better represented interaction types, leading to more prediction errors in the less represented interaction types [54].

Enhancers are short non-coding DNA fragments that assume an important part in the genetic process of gene expression and gene regulation (e.g., transcription and translation). Having high locational variation and free scattering in 98% of non-encoding genomes and due to their possibly distant location relative to the gene that is acted upon, enhancer identification is, therefore, more complicated than other genetic factors. There are many published works focused on identifying enhancers based on their sequence information, however, the resulting performance still requires improvements. To address this biological issue, several in silico studies have been done to identify and classify enhancer sequences among a myriad of DNA sequences using computational advances. Deep model ensembles hold the potential for improving on the best results achieved to date using shallow ML methods [55]. Although recent studies have come up with improved performance, shortfalls in these learning models still remain [56, 57].

Obstetric and Gynecological Applications

Introduction

Prediction and classification techniques have been well studied by ML researchers and developed for several real-word problems. However, the level of acceptance and success of prediction models are still below expectation due to some difficulties such as the low performance of prediction models when they are applied in different environments. Such a problem has been addressed by many researchers, mainly from the ML community. A second problem, principally raised by model users in different communities, such as managers, economists, engineers, biologists, and medical practitioners, etc., is the prediction models' interpretability. The latter is the ability of a model to explain its predictions and exhibit the causality relationships between the inputs and the outputs. In the case of classification, a successful way to alleviate the low performance is to use ensemble classifiers. It is an intuitive strategy to activate collaboration between different classifiers towards a better performance than individual classifiers. Unfortunately, ensemble classifier methods do not take into account the interpretability of the final classification outcome. It even worsens the original interpretability of the individual classifiers. Thus, ensemble classifiers constitute one of the leading research directions in ML and data mining. The use of multiple classifiers generally allows better predictive performance than that achievable with a single model. Several approaches exist in the literature that provides means to construct and aggregate such ensembles. However, these ensemble systems contain redundant members that, if removed, may further increase group diversity and produce better results. Smaller ensembles also relax the memory and storage requirements, reducing system's run-time overhead while improving overall efficiency [58]. An implementation of classifiers combination approach that does not only promote the overall performance but also preserves the interpretability of the resulting model has been validated with case studies from medical domain namely, obstetric and gynecological applications concerning mainly assisted reproductive technologies (ART), prenatal diagnosis, and gynecological oncology, problems where interpretability is critical to make appropriate clinical decisions [59].

Assisted Reproductive Technologies

In the past decades, the number of *in vitro* fertilization (IVF) procedures for the conception of a child has been rising continuously, however, the success rate of artificial insemination remained low. According to current statistics, large portion of unsuccessful IVF relates to some women's factors. As the directly related female organ, the proper investigation of the uterus has primary importance. Namely, visible markers may indicate inflammations or other negative effects that jeopardize successful implantation. The purpose of DL techniques is to support the observability of the uterus from this aspect by providing computer-aided tools for the extraction of its wall from video hysteroscopy. As for methodology, fully convolutional neural networks are used for the automatic segmentation of the video frames to determine the region of interest. Moreover, segmentation accuracy is increased by applying ensemble-based approaches at two levels. First, the predictions of a given fully convolutional neural network are aggregated for the overlapping regions of sub-images, which are derived from the splitting of the original images. Next, the segmentation results of different fully convolutional neural networks are fused via a weighted combination model; optimization for adjusting the weights are also provided. Based on these experimental results, 91.56% segmentation accuracy regarding the recognition of the uterus wall has been achieved [60]. Predicting the outcome of IVF treatment is an extremely semantic issue in reproductive medicine. Discrepancies in results among reproductive centers still exist making the construction of new systems capable to foresee the desired outcome a necessity. As such, ANN represent a combination of a learning, self-adapting, and predicting machine [61]. An ensemble of independently trained neural networks has been proposed for the assessment of luteinizing hormone (LH) surge for predicting ovulation time in infertile but ovulating women. The proposed ensemble involves a number of parallel neural network modules. Each pair of the neural networks learn specific data that are previously collected for monitoring timing function of LH estradiol (pg/ml), and follicle diameter (mm) is used to train neural network pairs to approximate the function of the LH values. A reasonable and accurate estimation places ovulation approximately 10-12 h after the LH peak. The double-valued (bi-phasic) regions of training data are separated into two single-valued (bi-phasic) parts (not exactly preovulatory/postovulatory division) that can be learned by each module of the neural network pair. During testing, after the initial decision to have single-valued sides, the assessment is obtained by a linear opinion pool (consensus rule) using the decisions of neural networks on the corresponding side without waiting. The network ensemble has various desirable properties: high assessment accuracy of a double-valued multisource data, minimized learning and recall times, and a parallel structure. The ovulation time can be predicted through the assessment of LH peak with better precision and fewer number of tests [62]. Sperm morphology, as an indicator of fertility, is a critical tool in semen analysis. A smartphone-based hybrid system that fully automates the sperm morphological analysis is introduced with the aim of eliminating unwanted human factors. Proposed hybrid system consists of two progressive steps: automatic segmentation of possible sperm shapes and classification of normal/abnormal sperms. In the segmentation step, clustering techniques with/without group sparsity approach were tested to extract the region of interests from the images. Subsequently, a novel publicly available morphological sperm image data set, whose labels were identified by experts as non-sperm, normal and abnormal sperm, was created as the ground truths of classification step. In the classification step, conventional and ensemble ML methods were applied to domain-specific features that were extracted by using wavelet transform and descriptors. Additionally, as an alternative to conventional features, three deep neural network architectures, which can extract high-level features from raw images after using statistical learning, were employed to increase the proposed method's performance. The results show that, for the conventional features, the highest classification accuracies were achieved as 80.5% and 83.8% by using the wavelet- and descriptor-based features that were fed to the SVM, respectively. On the other hand, the Mobile-Net, which is a very convenient network for smartphones, achieved 87% accuracy. In the light of obtained results, it is seen that a fully automatic hybrid system, which uses the group sparsity to enhance segmentation performance and the Mobile-Net to obtain high-level robust features, can be an effective mobile solution for the sperm morphology analysis problem. In conclusion, this is a fully automated hybrid human sperm detection and classification system based on mobile-net [63].

Embryonic Stem cells (ESC) can be differentiated into ectoderm, endoderm, and mesoderm derivatives, producing the majority of cell types. In regular culture conditions, ESC self-renewal is maintained through molecules that inhibit spontaneous differentiation enabling long-term cellular expansion. This undifferentiating condition is characterized by multiple metastable states that fluctuate between self-renewal and differentiation balance. A study aimed to characterize the high-pluripotent ESC meta-state marked by the expression of zinc finger and serum response element (SRE)-zip-code-binding *protein* (ZBP), C-terminal finger 51 (CTfin51), AW-1, and Number 18 complimentary-DNA (cDNA) (SCAN) domain-containing protein 4C (Zscan4) through a supervised ML framework based on an ensemble of SVM classifiers. This study revealed a leukemia inhibitor factor (LIF)-dependent not-canonical pluripotency signature (AF067063, BC061212, deubiquitinating enzyme 1 [Dub1], eukaryotic translation initiation factor 1A [Eif1a], Gm12794, Gm13871, Gm4340, Gm4850, mouse 2-cell-stage variable group members 1/3 [Tcstv1/3], and zinc finger protein 352 [Zfp352]), that specifically marks Zscan4 ESC fluctuation. This novel ESC meta-state is enhanced by high-pluripotency culture conditions obtained through extracellular signal regulated-kinase (ERK) and glycogen synthase kinase-3 (Gsk-3) signaling inhibition (2i). Significantly, it has been reported that the conditional ablation of the novel ESC meta-state marked by the expression of Gm12794 is required for ESC self-renewal maintenance. In conclusion, the comprehension of ESC biology has been extended through the identification of a novel molecular signature associated with pluripotency programming [64].

Prenatal Diagnosis

Central nervous system (CNS) anomalies represent a wide range of congenital birth defects, with an incidence of approximately 1% of all births. They are currently diagnosed using ultrasound evaluation. However, there is strong need for a more accurate and less operator-dependent screening method, such as a characterization of maternal serum in order to build a metabolomic fingerprint resulting from congenital anomalies of the CNS. In a case-control pilot study, metabolomic profiles were obtained from serum of 168 mothers (98 controls and 70 cases), using gas chromatography coupled to mass spectrometry. Nine ML and classification models were built and optimized. An ensemble model was built based on results from the individual models. All samples were randomly divided into two groups. One was used as training set, the other one for diagnostic performance assessment. Ensemble ML model correctly classified all cases and controls. Propanoic, lactic, gluconic, benzoic, oxalic, 2-hydroxy-3-methyl butyric ($C_5H_{10}O_3$), acetic, lauric, myristic, and stearic acid, myo-inositol, and mannose were selected as the most relevant metabolites in class separation. The metabolomic signature of second-trimester maternal serum from pregnancies affected by a fetal CNS anomaly is quantifiably different from that of a normal pregnancy. Maternal serum metabolomics is, therefore a promising tool for the accurate and sensitive screening of such congenital disabilities. Moreover, the details of the most relevant metabolites and their respective biochemical pathways allow better understanding of the overall pathophysiology of affected pregnancies [65]. MRI is a common imaging technique used extensively to study human brain activities. Recently, it has been used for scanning the fetal brain. Amongst 1000 pregnant women, 3 of them have fetuses with brain abnormality. Hence, the primary detection and classification are important. ML techniques have a large potential in aiding the early detection of these abnormalities, which correspondingly could enhance the diagnosis process and follow-up plans. Most research focused on the classification of abnormal brains in a primary age has been for newborns and premature infants, with fewer studies focusing on images for fetuses. These studies associated fetal scans to scans after birth for the detection and classification of brain defects early in the neonatal age. This type of brain abnormality is named "small for gestational age" (SGA). A novel framework for the classification of fetal brains at an early age (before the fetus is born) has been proposed. The study incorporates several ML classifiers, such as diagonal quadratic discriminates analysis, KNN, RF, naïve Bayes, and radial basis function neural network classifiers. Moreover, several bagging and Ada-boosting ensembles models have been constructed using RF, naïve Bayes, and radial basis function network classifiers. The performances of these ensembles have been compared with their individual models. The results show that this novel approach can successfully identify and classify numerous types of defects within MRI images of the fetal brain of various gestational ages. Using the KNN classifier, it was possible to achieve the highest classification accuracy and area under receiver operating characteristic (ROC) of 95.6% and 99%, respectively. In addition, ensemble classifiers improved the results of their respective individual models [66].

Stillbirth and Preterm Birth

Quantification of stillbirth risk has potential to support clinical decision-making. Studies that have attempted to quantify stillbirth risk have been hampered by small event rates, a limited range of predictors that typically exclude obstetric history, lack of validation, and restriction to a single classifier (logistic regression). Consequently, predictive performance remains low, and risk quantification has not been adopted into antenatal practice. The study population consisted of all births to women in Western Australia from 1980 to 2015, excluding terminations. After all exclusions, there were 947,025 livebirths and 5,788 stillbirths. Predictive models for stillbirth were developed using multiple ML classifiers: regularized logistic regression, DT based on classification and regression trees, RF, extreme gradient boosting, and a multilayer perceptron neural network. Ten-fold cross-validation using independent data not used to develop the models has been applied. Predictors included maternal socio-demographic characteristics, chronic medical conditions, obstetric complications, and family history in both the current and previous pregnancy. In this cohort, 66% of stillbirths were observed for multiparous women. The best performing classifier (extreme gradient boosting) predicted 45% (95% confidence interval [CI]: 43-46%) of stillbirths for all women and 45% (95% CI: 43-47%) of stillbirths after the inclusion of previous pregnancy history. Almost half of stillbirths could be potentially identified antenatally based on a combination of current pregnancy complications, congenital anomalies, maternal characteristics, and medical history. Greatest sensitivity is achieved with addition of current pregnancy complications. Ensemble classifiers offered marginal improvement for prediction compared to logistic regression [67]. High rate of preterm birth (birth before 37 weeks of gestation) in the world and its adverse outcomes for pregnant women and newborns necessitate to predict preterm birth and identify its main risk factors. Premature deliveries have been divided into provider-initiated (with medical intervention for early terminating the pregnancy) and spontaneous preterm birth (without any intervention) categories in the previous studies. A study was conducted with the main aim of proposing methods for prediction of provider-initiated preterm birth and spontaneous premature deliveries and ranking the predictive features. Data from the Iranian databank of maternal and neonatal records (IMAN registry) was used in the study. The collected data had information about more than 1,400,000 deliveries with 112 features. Among them, 116,080 preterm births have occurred (from which 11,799 and 104,281 cases belong to provider-initiated preterm birth and spontaneous premature delivery, respectively). The data can be considered as big data due to its large number of data records, a large number of features, and unbalanced distribution of the data between three classes of term, provider-initiated, and spontaneous preterm birth. Map Reduce-based ML algorithms have been proposed for this purpose. Map phase uses parallel feature selection and classification methods to score the features. Reduce phase aggregates the feature scores obtained in Map phase and assign final scores to the features. Moreover, the classifiers trained in Map phase are aggregated based on two different ensemble rules in Reduce phase. Experimental results show that the best performance of the proposed models for preterm birth prediction is accuracy of 81% and the area under

ROC of 68%. Top features for predicting term, provider-initiated preterm, and spontaneous premature birth identified in this study were having pregnancy risk factors, having gestational diabetes, having cardiovascular disease, maternal underlying diseases, and mother age. Chronic blood pressure is a high rank feature for preterm birth prediction and father nationality is highly important for discriminating provider-initiated from spontaneous premature delivery. Identifying the pregnant women with high risk of spontaneous premature or therapeutic preterm delivery in the proposed model can help them to:

1) reduce the probability of premature birth with monitoring and management of the main risk factors; and/or
2) educate them to care from the premature newborn.

Management and monitoring top features the discriminating term, provider-initiated preterm, and spontaneous premature birth or their associated factors can reduce preterm labor or its negative outcomes [68].

Nutrition

Nutritional phenotyping can help achieve personalized nutrition, and ML tools may offer novel means to achieve phenotyping. A study was conducted with the primary aim to use energy balance components, namely input (dietary energy intake and macronutrient composition) and output (physical activity), to predict energy stores (body weight) as a way to evaluate their ability to identify potential phenotypes based on these parameters. From the Women's Health Initiative (WHI) Observational Study, carbohydrates, proteins, fats, fibers, sugars, and physical activity variables, namely energy expended from mild, moderate, and vigorous-intensity activity, were used to predict current body weight (both as body weight in kilograms and as a body mass index [BMI] category). Several ML tools were used for this prediction. Finally, cluster analysis was used to identify putative phenotypes. For the numerical predictions, SVM, neural network, and KNN algorithms performed modestly, with mean approximate errors (MAE) of 6.70 kg, 6.98 kg, and 6.90 kg, respectively. For categorical prediction, SVM performed the best (54.5% accuracy), followed closely by the bagged tree ensemble and KNN algorithms. K-means cluster analysis improved prediction using numerical data, identified 10 clusters suggestive of phenotypes, with a minimum MAE of ~1.1 kg. A classifier was used to phenotype subjects into the identified clusters, with MAE <5 kg for 15% of the test set (n = ~2000). This study highlighted the challenges, limitations, and successes in using ML tools on self-reported data to identify determinants of energy balance [69].

Gynecological Oncology

Nowadays, the most frequent cancer in women is breast cancer (malignant tumor). Breast cancer also has the second highest frequency of death rate among women worldwide. Early-stage prevention becomes complex due to reasons unknown. In fact, if breast cancer is detected at the beginning stage, it can often be cured. Many researchers proposed numerous methods for early prediction of this cancer. A research combined the feature selection and optimization algorithms as the optimal tool to improve the learning performance like predictive accuracy of the Wisconsin Diagnostic Breast Cancer (WDBC) dataset classification from the UCI ML repository. An ensemble of the reduced data patterns based on feature selection was used to train a neural network using the Levenberg-Marquardt and the Particle Swarm Optimization algorithms to devise the appropriate neural network training weighting parameters, and then construct an effective neural network classifier to improve the WDBC classification's accuracy and efficiency. Experimental results show that the accuracy and the area under ROC improved emphatically, and the best performance in accuracy and ROC were 98.83% and 0.9971, respectively [70]. A feature EL based on Sparse Autoencoders and Soft-max Regression for classification of breast cancer into benign (non-cancerous) and malignant (cancerous) has been proposed. It used Wisconsin Diagnostic Breast Cancer (WDBC) dataset. The proposed method is assessed using various performance indices like true classification accuracy, specificity, sensitivity, recall, precision, *f* measure, and multi-case-control study on cancer. Simulation and result proved that the proposed approach gives better results in terms of different parameters. The prediction results obtained by the proposed approach were very promising (98.60% exact accuracy). In addition, the proposed method outperforms the Stacked Sparse Autoencoders and Soft-max Regression-based model and other state-of-the-art classifiers in terms of various performance indices. Experimental simulations, empirical results, and statistical analyses are also showing that the proposed model is an efficient and beneficial model for the classification of breast cancer. It is also comparable with the existing ML and soft computing approaches present in the related literature [71].

Mammography (X-ray images of the breast) screening currently relies on subjective human interpretation. AI advances could be used to increase mammography screening accuracy by reducing missed cancers and false positives. A study aimed to evaluate whether AI can overcome human mammography interpretation limitations with a rigorous, unbiased evaluation of ML algorithms. In this diagnostic accuracy study, an international, crowdsourced challenge was hosted to foster AI algorithm development focused on interpreting screening mammography. More than 1100 participants comprising 126 teams from 44 countries participated. Algorithms used images alone (challenge 1) or combined images, previous examinations (if available), and clinical and demographic risk factor data (challenge 2), and output a score that translated to cancer yes/no within 12 months. Algorithm accuracy for breast cancer detection was evaluated using area under the curve (AUC) and algorithm specificity compared with radiologists' specificity with radiologists' sensitivity set at 85.9% (US) and 83.9% (Sweden). An ensemble method aggregating top-performing AI algorithms and radiologists' recall assessment was developed and evaluated. Overall, 144,231 screening mammograms from 85,580 US women (952 cancer positive ≤12 months from screening) were used for algorithm training and validation. A second independent validation cohort included 166,578 examinations from 68,008 Swedish women (780 cancer positive). The top-performing algorithm achieved an AUC of 0.858 (US) and 0.903 (Sweden), and 66.2% (US) and 81.2% (Sweden) specificity at the radiologists' sensitivity, lower than community-practice radiologists' specificity of 90.5% (US) and 98.5% (Sweden). Combining top-performing algorithms and US radiologist assessments resulted in a higher AUC of 0.942 and achieved a significantly improved specificity (92.0%) at the same sensitivity. While no single AI algorithm outperformed radiologists, an ensemble of AI algorithms combined with radiologist assessment in a single-reader screening environment improved overall accuracy. This study underscored the potential of using ML methods for enhancing mammography screening interpretation [72]. Some typical signatures like masses and microcalcifications upon investigating mammograms can help diagnose women better. Manual diagnosis is a hard task the radiologists carry out frequently. For their assistance, many CAD approaches have been developed. Several state-of-the-art ML methods for automated classification of clustered microcalcifications have been investigated. A system that has been used in a study with more than 60,000 patients is CAD for the early assisted diagnosis of breast cancer. On two mammograms of two different patients taken from this study, the radiologist revealed a suspicious lesion, subsequently revealed to be malignant tumors at an early stage by biopsy. It is reasonable to think that the life of these two patients was saved thanks to this diagnostic examination, and above all thanks to the experience of the radiologist and other physicians who intervened subsequently. On the same images, the suspect region was automatically outlined by a CAD system based on ML (specifically, boosting algorithms). We can see the almost perfect matching between man and machine: both have found the same diagnostic sign. Since the system is in good agreement with the radiologist, it becomes then a good candidate to help him in the diagnosis, where the radiologist has a doubt or more simply is fatigued by the number of images to be examined at the end of the day [74, 74]. Thanks to the radiologist-CAD collaboration (based on ML), the number of lives saved can increase [75]. To improve upon the state-of-the-art, a deep ensemble transfer learning and neural network classifier for automatic feature extraction and classification has been proposed. In computer-assisted mammography, DL-based architectures are generally not trained on mammogram images directly. Instead, the images are pre-processed beforehand, and then they are adopted to be given as input to the proposed ensemble model. The robust features extracted from the ensemble model were optimized into a feature vector and were further classified using the neural network (nntraintool). The network was trained and tested to separate benign and malignant tumors, thus achieving an accuracy of 0.88 with an AUC of 0.88. The attained results show that the proposed methodology is a promising and robust CAD system for breast cancer classification [76]. The classifier is part of a CAD scheme that is aimed to assisting radiologists in making more accurate diagnoses of breast cancer on mammograms. The considered methods are: SVM, kernel Fisher discriminant (KFD), relevance vector machine (RVM), and committee machines (ensemble averaging and AdaBoost), of which most have been developed recently in statistical learning theory. Differentiation of malignant from benign microcalcifications has been formulated as a supervised learning problem, and these learning methods have been applied to develop the classification algorithm. As input, these methods used image features automatically extracted from clustered microcalcifications. These methods have been tested using a database of 697 clinical mammograms from 386 cases, which included a wide spectrum of difficult-to-classify cases. The distribution of the cases in this database has been analyzed using the multidimensional scaling technique, which reveals that in the feature space, the malignant cases are not trivially separable from the benign ones. ROC analysis has been used to evaluate and to compare classification performance by the different methods. In addition, it has also been investigated how to combine information from multiple-view mammograms of the same case, so that classifier can make the best decision. In these experiments, the kernel-based methods (i.e., SVM, KFD, and RVM) yielded the best performance (Az = 0.85, SVM), significantly outperforming a well-established, clinically-proven CAD approach that is based on neural network (Az = 0.80) [77]. Accurate classifiers are vital to design precise CAD systems. Classification performances of ML algorithms are sensitive to data characteristics. In this aspect, determining the relevant and discriminative features is a key step to improve performance of CAD. There are various feature extraction methods in the literature. However, there is no universal variable selection algorithm that performs well in every data analysis scheme. RF, an ensemble of trees, is used in classification studies successfully. The success of RF algorithms makes it eligible to be used as kernel of a wrapper feature subset evaluator. The experimental results demonstrated the efficiency of proposed feature extraction strategy with the increase in most of the classification accuracies of the algorithms [78]. The purpose of a study was to develop a CAD system for the classification of malignant and benign masses in the breast using ultrasonography based on a convolutional neural network, a state-of-the-art DL technique. The regions for the correct classification have been explored by generating a heat map that presented the important regions used by the convolutional neural network for human malignancy/benign classification. Clinical data was obtained from a large-scale clinical trial previously conducted by the Japan Association of Breast and Thyroid Sonology (JABTS). Images of 1536 breast masses (897 malignant and 639 benign) confirmed by pathological examinations were collected, with each breast mass captured from various angles using an ultrasound imaging probe. An ensemble network has been constructed by combining two convolutional neural network models (Visual Geometry Group 19 [VGG19] and Residual Network 152 [ResNet152]) fine-tuned on balanced training data with augmentation and used the mass-level classification method to enable the convolutional neural network to classify a given mass using all views. For an independent test set consisting of 154 masses (77 malignant and 77 benign), this network showed outstanding classification performance with a sensitivity of 90.9% (95% CI: 84.5-97.3%), a specificity of 87.0% (79.5-94.5%), and AUC of 0.951 (0.916-0.987) compared to that of the two convolutional neural network models. In addition, this study indicated that breast masses themselves were not detected by the convolutional neural network as important regions for correct mass classification. Collectively, this convolutional neural network-based CAD system is expected to assist physicians by improving the diagnosis of breast cancer in clinical practice [79].

Microwave-based breast cancer detection has been proposed as a complementary approach to compensate for some drawbacks of existing breast cancer detection techniques. Among the existing microwave breast cancer detection methods, ML-type algorithms have recently become more popular. These focus on detecting the existence of breast tumors rather than performing imaging to identify the exact tumor position. A vital component of the ML approaches is feature extraction. One of the most widely used feature extraction method is principal component analysis. However, it can be sensitive to signal misalignment. Feature extraction methods based on time-frequency representations of microwave data include the wavelet transform and the empirical mode decomposition. Time-invariant statistics can be generated to provide more robust features to data misalignment. Experimental results show that features extracted from decomposition results of the wavelet transform and empirical mode decomposition improve the detection performance when combined with an ensemble selection-based classifier [80]. In the classification of breast cancer from medical diagnostic data, information gain has been adapted for feature selections. Neural fuzzy, KNN, and quadratic classifier, each single model scheme as well as their associated ensemble ones, have been developed for classifications. In addition, a combined ensemble model with these three schemes has been constructed for further validations. The experimental results indicate that EL performs better than individual single ones. Moreover, the combined ensemble model illustrates the highest accuracy of classifications for breast cancer among all models [81].

Despite existing prognostic markers, breast cancer prognosis remains a difficult subject due to the complex relationships between many contributing factors and survival. Current breast cancer predictive signatures are not unique. From the ML perspective, it is well known that combining multiple classifiers can improve classification performance. An ensemble ML approach, which consists of choosing feature subsets and learning predictive models from them, has been proposed. This method significantly improves classification performance with a particular emphasis on sensitivity, which is critical to avoid misclassifying poor prognosis patients as good prognosis [82]. A study seeked to integrate multiple clinicopathological and genomic factors with dimensional reduction across ML algorithms to compare survival predictions. In a secondary analysis of the data from a prospective cohort study of female patients with breast cancer enrolled in the Molecular Taxonomy of Breast Cancer International Consortium (METABRIC), a series of predictive models has been constructed: ensemble models (gradient boosting and RF), SVM, and ANN for 5-year survival based on clinicopathological and gene expression data after K-means clustering with KNN classification. Model performance was evaluated by ROC curve, accuracy, and calibration slope. Model stability was assessed over 10 random runs in terms of ROC, accuracy, calibration slope, and variable importance. The analytic cohort was composed of 1874 patients with breast cancer. Overall, the median age was 62 years; the 5-year survival rate was 75%. ROC and accuracy were not significantly different between models (ROC and accuracy around 0.67 and 0.72 across models, respectively). However, ensemble methods resulted in better fit (calibration slope) with stable measures of variable importance across ten random training/validation splits. K-means clustering of gene expression profiles on training data points along with KNN classification of validation data points was a robust method of dimensional reduction. Furthermore, the gene expression cluster with the highest mortality risk was an influential factor in model prediction. Using ML methods to construct predictive models for 5-year survival in patients with breast cancer, discrimination ability across models has been demonstrated with new insight into the stability and utility of dimensional reduction on genomic features in breast cancer survival prediction [83]. Wavelet neural networks are a combination of neural networks and wavelets and have been mostly used in the area of time-series prediction and control. Recently, evolutionary wavelet neural networks have been employed to develop cancer prediction models. A study has proposed to use ensembles of evolutionary wavelet neural networks. The search for a high-quality ensemble was directed by a fitness function that incorporates the accuracy of the classifiers both independently and as part of the ensemble itself. The ensemble approach was tested on one publicly available biomedical benchmark datasets on breast cancer, using a 10-fold cross-validation strategy. These experimental results show that, for the dataset, the performance was similar to previous studies reported in the literature [84]. A unified and flexible framework for EL in the presence of censoring has been proposed. For right-censored data, a RF algorithm and a generic gradient boosting algorithm for the construction of prognostic and diagnostic models have been introduced. The diagnostic capabilities of the proposed censored data RF and boosting methods can be applied to the recurrence-free survival (RFS) time of node-positive breast cancer patients [85].

Biomarkers are a key component of precision medicine. However, full clinical integration of biomarkers has been met with challenges, partly attributed to analytical difficulties. It has been shown that biomarker reproducibility is susceptible to data preprocessing approaches. ML ensembles of preprocessing methods are a general strategy to improve biomarker performance for prediction of survival from early breast cancer. Breast cancer patients have been stratified into either low-risk or high-risk groups based on four published hypoxia signatures, using 24 different preprocessing approaches for microarray normalization. The 24 binary risk profiles determined for each hypoxia signature were combined using a RF to evaluate the efficacy of a preprocessing ensemble classifier. It has been demonstrated that the best way of merging preprocessing methods varies from signature to signature, and that there is likely no "best" preprocessing pipeline that is universal across datasets, highlighting the need to evaluate ensembles of preprocessing algorithms. Further, novel signatures for each preprocessing method have been developed, and the risk classifications from each were incorporated in a meta-RF model. Interestingly, the classification of these biomarkers and its ensemble show striking consistency, demonstrating that similar intrinsic biological information is being faithfully represented. As such, these classification patterns further confirm that there is a subset of patients whose prognosis is consistently challenging to predict. Performance of different prognostic signatures varies with the preprocessing method. A simple classifier by unanimous voting of classifications is a reliable way of improving on single preprocessing methods. Future signatures will likely require integration of intrinsic and extrinsic clinicopathological variables to better predict disease-related outcomes [86]. The aim of a study was to explore new salivary biomarkers to discriminate breast cancer patients from healthy controls. Saliva samples were collected after 9 h fasting and were immediately stored at -80°C. Capillary electrophoresis and liquid chromatography with mass spectrometry were used to quantify hundreds of hydrophilic metabolites. Conventional statistical analyses and AI-based methods were used to assess the discrimination abilities of the quantified metabolites. A multiple logistic regression (MLR) model and an alternative DT-based ML method were used. The generalization abilities of these mathematical models were validated in various computational tests, such as cross-validation and resampling methods. One hundred sixty-six unstimulated saliva samples were collected from 101 patients with invasive carcinoma of the breast, 23 patients with ductal carcinoma in situ (DCIS), and 42 healthy controls. Of the 260 quantified metabolites, polyamines were significantly elevated in the saliva of patients with breast cancer. Spermine showed the highest area under the ROC [0.766; 95% CI: 0.671-0.840; $p < 0.0001$] to discriminate patients with invasive carcinoma from controls. In addition to spermine, polyamines and their acetylated forms were elevated in patients with invasive carcinoma only. Two hundred each of two-fold, five-fold, and ten-fold cross-validation using different random values were conducted and the MLR model had slightly better accuracy. The alternative DT with an ensemble approach showed higher accuracy (0.912; 95% CI: 0.838-0.961; $p < 0.0001$). These prediction models also included spermine as a predictive factor. These data indicated that combinations of salivary metabolomics with the alternative DT-based ML methods show potential for non-invasive screening of breast cancer [87].

Learning accurate models from 'omics data is bringing many challenges due to their inherent high-dimensionality, e.g., the number of gene expression variables, and comparatively lower sample sizes, which leads to ill-posed inverse problems. Furthermore, the presence of outliers, either experimental errors or interesting, abnormal clinical cases, may severely hamper a correct classification of patients and the identification of reliable biomarkers for a particular disease. It has been proposed to address this problem through an ensemble classification setting based on distinct feature selection and modeling strategies, including logistic regression with elastic net regularization, Sparse Partial Least Squares-Discriminant Analysis

(SPLS-DA) and Sparse Generalized PLS (SGPLS), coupled with an evaluation of the individuals' outlier-ness based on the Cook's distance. The consensus is achieved with the Rank Product statistics corrected for multiple testing, which gives a final list of sorted observations by their outlier-ness level. This strategy has been applied for the classification of Triple-Negative Breast Cancer (TNBC) RNA-Seq and clinical data from the Cancer Genome Atlas (CGA). The detected 24 outliers were identified as putative mislabeled samples, corresponding to individuals with discrepant clinical labels for the human epidermal growth factor 2 (HER2) receptor, but also individuals with abnormal expression values of estrogen receptor (ER), progesterone receptor (PR), and HER2, contradictory with the corresponding clinical labels, which may invalidate the initial TNBC label. Moreover, the model consensus approach leads to the selection of a set of genes that may be linked to the disease. These results are robust to a resampling approach, either by selecting a subset of patients or a subset of genes, with a significant overlap of the outlier patients identified. The proposed ensemble outlier detection approach constitutes a robust procedure to identify abnormal cases and consensus covariates, which may improve biomarker selection for precision medicine applications. The method can also be easily extended to other regression models and datasets [88]. ERα plays a significant role in the occurrence of breast cancer and may cause various adverse side-effects when ERα is an off-target protein. A theoretical model was derived to predict the binding affinity of ERα using the pharmacophore ensemble/SVM scheme to consider the promiscuous characteristic of ERα. The estimations by pharmacophore ensemble/SVM were discovered to be in good agreement with the observed values for those training molecules ($n = 31$, $r2 = 0.80$, $qCV2 = 0.77$, $RMSE = 0.57$, and $s = 0.58$), test molecules ($n = 179$, $q2 = 0.91$-0.96, $RMSE = 0.33$, and $s = 0.26$), and outliers ($n = 15$, $q2 = 0.80$-0.86, $RMSE = 0.56$, and $s = 0.49$). When subjected to various statistical validations, the pharmacophore ensemble/SVM model consistently fulfilled the strictest criteria. A mock test also asserted its predictivity. When compared with crystal structures, the calculated results are consistent with the reported ERα-ligand co-complex structure, and the plasticity nature of ERα is also disclosed. Consequently, this precise, fast, and robust model can be adopted to predict ERα-ligand binding affinities and to design safer non-ERα-targeted pharmaceuticals in the process of drug discovery and development [89].

Epithelial ovarian cancer is the deadliest cancer of the female reproductive system. Ovarian tumors are the most common diagnostic challenge for gynecologists, and ultrasound examination has become the primary technique for assessment of ovarian pathology and for preoperative distinction between malignant and benign ovarian tumors. However, ultrasonography is highly examiner-dependent, and there may be an important variability between two different specialists when examining the same case. The objective of a study was the evaluation of different well-known ML systems to perform the automatic categorization of ovarian tumors from ultrasound images. A real patient database whose input features have been extracted from 348 images, from the International Ovarian *Tumor* Analysis (IOTA) tumor images database, holding together with the class labels of the images, has been used. For each patient case and ultrasound image, its input features have been previously extracted using Fourier descriptors computed on the Region of Interest (ROI). Then, four ML techniques were considered for performing the classification stage: KNN, Linear Discriminant, SVM, and Extreme Learning Machine. According to the obtained results, the KNN classifier provides inaccurate predictions (less than 60% of accuracy) independently of the size of the local approximation. In contrast, the classifiers based on Linear Discriminant, SVM, and Extreme Learning Machine are robust in this biomedical classification (more than 85% of accuracy). ML methods can be efficiently used for developing the classification stage in CAD systems of ovarian tumor from ultrasound images. These approaches are able to provide automatic classification with a high rate of accuracy. Future work should aim at enhancing the classifier design using ensemble techniques. Another ongoing work is to exploit different kind of features extracted from ultrasound images [90]. To date, there is no effective screening method for early detection of ovarian cancer and current diagnostic armamentarium may include sonographic grading of the tumor and analyzing serum levels of tumor markers, cancer antigen 125 (CA-125) and human epididymis protein 4 (HE4). Microorganisms (bacterial, archaeal, and fungal cells) residing in mucosal tissues including the gastrointestinal and urogenital tracts can be altered by different disease states, and these shifts in microbial dynamics may help to diagnose disease states. It has been hypothesized that the peritoneal microbial environment was altered in patients with ovarian cancer and that the inclusion of selected peritoneal microbial features with current clinical features into prediction analyses will improve detection accuracy of patients with ovarian cancer. Blood and peritoneal fluid were collected from consented patients that had sonography confirmed adnexal masses and were being seen at Southern Illinois University (SIU) School of Medicine Simmons Cancer Institute. Blood was processed and serum HE4 and CA-125 were measured. Peritoneal fluid was collected at the time of surgery and processed for next generation sequencing (NGS) using 16S V4 exon bacterial primers and bioinformatics analyses. It has been found that patients with ovarian cancer had a unique peritoneal microbial profile compared to patients with a benign mass. Using ensemble modeling and ML pathways, 18 microbial features that were highly specific to ovarian cancer pathology were identified. Prediction analyses confirmed that inclusion of microbial features with serum tumor marker levels and control features (patient age and BMI) improved diagnostic accuracy compared to currently used models. The conclusion was that ovarian cancer pathogenesis alters the peritoneal microbial environment and that these unique microbial features are important for accurate diagnosis of ovarian cancer. This study warrants further analyses of the importance of microbial features in regards to oncological diagnostics and possible prognostic and interventional medicine [91].

Endometrial cancer is the most common cancer of the female reproductive tract in developed countries. At the moment, no effective screening system is available. A study has evaluated the diagnostic performance of a serum metabolomic signature. Two enrollments were carried out: one consisting of 168 subjects (88 with endometrial cancer and 80 healthy women) was used for building the classification models, the second (used to establish the performance of the classification algorithm) consisted of 120 subjects (30 with endometrial cancer, 30 with ovarian cancer, 10 with benign endometrial disease, and 50 healthy controls). Two ensemble models were built, one with all endometrial cancer patients versus controls (Model I) and one in which endometrial cancer patients were aggregated according to their histotype (Model II). Serum metabolomic analysis was conducted via gas chromatography-mass spectrometry, while classification was done by an EL machine. Accuracy ranged from 62% to 99% for Model I and from 67% to 100% for Model II. The ensemble model showed an accuracy of 100% both for Model I and II. The most critical metabolites in class separation were lactic acid, progesterone, homocysteine, 3-hydroxybutyrate, linoleic acid, stearic acid, myristic acid, threonine, and valine. The serum metabolomics signature of endometrial cancer patients is peculiar because it differs from that of healthy controls and from that of benign endometrial disease and from other gynecological cancers (such as ovarian cancer) [92]. A study aimed to build DL-based radiomic methods in differentiating vessel invasion from non-vessel invasion in cervical cancer with multi-parametric MRI data. A set of 1070 dynamic T1 contrast-enhanced (DCE-T1) and 986 T2 weighted imaging (T2WI) MRI images from 167 early-stage cervical cancer patients were used to train and validate DL models. Predictive performances were evaluated using ROC curve and confusion matrix analysis, with the DCE-T1 showing more discriminative results than T2WI MRI. By adopting an attention EL strategy that integrates both MRI sequences, the highest average area was obtained under the ROC of 0.911 (sensitivity = 0.881 and specificity = 0.752). The superior performances in this study, when compared to existing radiomic methods, indicate that a wealth of DL-based radiomics could be developed to aid radiologists in preoperatively predicting vessel invasion in cervical cancer patients [93].

LIGHTS AND SHADOWS

In the last ten years, there has been a significant and growing use of AI in Medicine, which is raising great interest also thanks to publications that have revealed a particular precision in specific clinical contexts. In particular, thanks to DL, some computer systems have managed to have a diagnostic accuracy comparable to that of physicians experts in various disciplines. However, to date, there is still a lack of broad-spectrum studies and applications on its effectiveness on important clinical goals such as reducing mortality or improving patient's life. Numerous advantages can be expected from the application of ML to Medicine: from increased system productivity to higher diagnostic accuracy and prediction of epidemic events, up to the possibility to encourage access to diagnostic tests even in places and to people who do not take advantage of them because of geographic, political, and economic barriers. For example, some ML health applications that lead to an improvement in diagnostic accuracy are:

1) identification of tumors through the analysis of radiological images;
2) detection of diabetic retinopathy; and
3) algorithms capable of predicting future cardiovascular events.

Furthermore, even if it has no relevance to clinical purposes, the introduction of chatbots and vocal assistants in the relationship with the public could improve the support to patients. Usually, the ML approach that is used in Medicine is the supervised one; the input is the diagnostic image that is provided or the description of a clinical case, and the output is the corresponding prediction. The output is typically expressed as a category, a score, or a percentage value. In the training phase, a large number of cases, already classified according to the "gold standard," i.e., a diagnosis defined by a group of specialists, are presented to the system. Subsequently, in the test phase, a series of images are shown, always classified by experts, but without the model being aware of the classification. At this stage, its predictive ability and its diagnostic accuracy with respect to cases already correctly classified are observed. This phase is repeated until levels of satisfactory accuracy are reached. Some systems used in today's medical practice are shown below as an example:

1) Google's DeepMind Health is capable of analyzing millions of medical information in a few minutes, thus speeding up health processes like archiving medical records [94];
2) International Business Machines Corporation (IBM)'s Watson is a system that can answer questions made in natural language, used as a support system for clinical decisions [95];
3) a Stanford University's study reports how a specific algorithm succeeds, based on little information on the chemical structure of a drug, to formulate predictions on both toxicity and instability of the molecule, accelerating the synthetizing time of the drug [96]; and
4) a DL algorithm that can detect skin cancer with equal or better accuracy than that of doctors was developed at Stanford University [97].

There are also some specific apps that support the patient in daily life:

1) *Molly*: the first virtual nurse in the world developed with the goal to help patients by monitoring their health and the status of ongoing treatments [98]; and
2) *AiCure*: it uses the smartphone webcam to confirm in an autonomous way that the patient is in line with the indicated treatments [99].

These apps, like EHR itself, have been widely adopted at many healthcare facilities in an attempt to improve the quality of patient care and increase the productivity and efficiency of healthcare delivery. These EHR can accurately diagnose diseases, if utilized appropriately. While EHR can potentially resolve many of the existing problems associated with disease diagnosis, one of the main obstacles in effectively using them is the patient privacy and sensitivity of the medical information available in the EHR. Due to these concerns, even if EHR is available for storage and retrieval purposes, sharing of the patient records between different healthcare facilities has become a major concern and has hampered some of the practical advantages of using EHR. Due to this lack of data sharing, most of the facilities aim at building clinical decision support systems using limited amount of patient data from their own EHR systems to provide important diagnosis-related decisions. It becomes quite infeasible for a newly established healthcare facility to build a robust decision-making system due to the lack of sufficient patient records. However, to make effective decisions from clinical data, it is indispensable to have large amounts of data to train the decision models. In this regard, there are conflicting objectives of preserving patient privacy and having sufficient data for modeling and decision making [100].

Although the use of AI/ML in Medicine has already brought numerous positive results and, as previously expressed, there are great expectations about its future use, there are still concerns about these innovative technologies to support clinical-health activity. The main negative aspects regarding the use of AI/ML in Medicine are listed below:

1) one of the main criticisms is that of "over-reliance," namely that over time, physicians develop excessive trust in the skills of automation at the expense of their experience. Trust would be fueled by the perception that every new technology is better than the previous one, without a real evaluation of its effectiveness. The consequence of this excessive trust in the decision-making support supplied by AI/ML could lead to a dependence on the use of these systems (overdependence), which over time would bring to the deskilling of medical personnel;
2) during the training phase of the algorithms, images chosen by physicians, which have already been considered as correct, are subjected to the system, that is, images previously classified by specialists. This approach presents two potential problems:
   a) the possible difference between the quality of the images used in the training phase compared to those present in the clinical records, under which ML systems will have to produce their predictions. The quality of the images used in the training phase is high, since they have undergone a cleaning and pre-processing process that would be impossible to sustain in daily clinical practice. The quality of the so-called "real-word data"

rarely meets the "ideal data" assumptions that should feed an ML algorithm; and

b) the possibility that there is more than one "gold standard" (a sample test used as a reference to evaluate other tests) in the same clinical case, and each would lead to different predictions;

3) the difficulty to attributing medico-legal responsibilities in the event the physician decided to use the support of ML systems, which could lead him to errors; or the attribution of responsibility in case the physician decides not to apply the recommendations of the same systems;
4) in the history of Medicine, even recently, the introduction of new diagnostic or therapeutic instruments not sufficiently supported by scientific evidences has led, on several occasions, to the withdrawal of these instruments because of their ineffectiveness, or even harm, highlighted by the unfortunately late execution of well-conducted clinical studies, according to the phenomenon that has been called "medical reversal"; and
5) a further criticism is represented by the concept of uncertainty. An example may be the one cited by a recently published study by Dharmarajan et al. focused on elderly hospitalized patients for acute cardiopulmonary pathologies to which, upon access to hospital, a heart failure, a chronic obstructive pulmonary disease, or a pneumonia diagnosis was made. The study observed that during the hospital stay, patients received regular medical treatment for two or more of the above morbid conditions at the same time, and therefore not only for the primary diagnosis carried out at the time of admission. This study exemplifies how in real clinical practice, clinical pictures are often placed in "gray areas" and are not easily associable with "golden" diagnostic criteria as reported in medical texts or guidelines [101].

So, despite AI/ML systems are progressively spreading successfully in the field of Medicine in several clinical disciplines, there are still perplexities and criticisms on a stronger adoption of these innovative technologies. The potential barriers to the practical and massive introduction of AI/ML to support clinical decisions derives from the very nature of medical science, which is not deterministic, but always and still subject to subjective interpretation and evaluation by the physician based on his experience. The first problem is the availability of valid digital data to the algorithms because a large volume of information does not automatically correspond to a better quality of inferences and applications that derive from them. The data itself is unnecessary. In order for them to be truly useful, they must be selected, structured, and interpreted [102]. It is therefore, not the technologies that are decisive but the ability to extract value from their use. The data is not a closed entity, "data," but a social construct, the concrete result of specific cultural, social, technical, and economic choices made by individuals, institutions, or companies to collect, analyze, and use information and knowledge. The high expectations for AI also risk underestimating the dangers connected to an uncritical acceptance of related technologies, in particular, decision-making systems based on DL, algorithms that are not able to follow predefined rules but to "learn" from data itself, autonomously, detecting "hidden" patterns, which often even the experts are unable to explain, so much that the operating mode of these systems has been defined as a "black box" [103]. The necessary data for the training of algorithms to develop predictive models are generally of less than optimal quality, because they are not subjected to that process of "cleaning up" and reworking which would be unsustainable in daily clinical practice and therefore may not be able to provide implementable answers for clinical decisions and treatments, also because, sometimes, they could "learn" the errors of natural intelligence [104]. In addition to being unstructured, the data of the so-called "real world," for example EHR, are not always available, they are not available everywhere, and their value is also limited to the collection setting. The contextual aspects, difficult to explain in quantitative terms, can therefore, be underestimated and under-represented, just think of the conditions that are not well defined in terms of pathology, the "fragility," the conditions of extra-clinical discomfort, psychological, social, family factors, and conditions of economic or cultural, logistical disadvantage, which always influence the patient's clinical management. For example, clinical data alone have limited predictive power in the case of patients whose risk of re-hospitalization is mainly dependent on social determinants [105]. Another limit to the reliability of data is linked to uncertainty, an inevitable variable in Medicine, characterized by large gray areas of knowledge, due to the incomplete domain of available knowledge and the intrinsic limits of medical knowledge. In the case of studies on new diagnostic tests, for example, the reference "gold standard" can be multiple and in any case, subject to the risk of uncertainty in interpretation. Depending on the choice of which to use for algorithm training, the levels of accuracy may be different. Another critical issue concerns the predictive capabilities of AI in the prognostic field. In fact, the use of big data has limits that depend on their very nature: risks of bias in the sample selection, and in the collection and interpretation of the information that is processed, capable of threatening the validity and generalizability of the conclusions. In the epidemiological field, for example, data flows are particularly useful as they allow us to take a photograph, often instantaneous, of a particular phenomenon at a given moment, but do not allow us to grasp the aspects related to the interactions between citizens/patients and the context, often difficult to represent and explicit expression in terms of digitization, due to the possible presence of confounding variables and spurious correlations. Validation by external epidemiological sources is therefore necessary, in order not to arrive at wrong causal inferences that could lead to the removal of resources from interventions of proven efficacy [106].

AI is changing the cultural paradigm of Medicine: its applications may become increasingly indispensable to provide answers in contexts of high complexity and uncertainty and allow physicians to have more time to take care of the needs of their patient. The data, however, are not values, any intervention based on them must be customized, also taking into account the frequent contradiction of the knowledge provided by the literature. AI will be useful essentially as complementary to the physician, who will be able to delegate the calculations and operations on data but keep the interpretation of complex phenomena and the consequent possible solutions to themselves. AI systems must be considered an instrument, such as a microscope, a stethoscope, and an electrocardiograph, developed over time to make up for the limited perceptive capacity of physicians. The best results are expected when AI works in support of health workers, "second set of eyes," a tool of cultural integration between humans and smart machines, avoiding emphasizing disputes, basically quite irrelevant, on which cognitive system, human or artificial, is more "intelligent" [107]. As stated by A. Verghese, "clinicians should seek an alliance in which machines predict (with significantly greater accuracy), and humans explain, decide, and act" [108]. Physicians must, therefore, play a leading, supervising, and monitoring role, using their intelligence and the skills that make them superior to machines, in particular abstraction, intuition, flexibility, and empathy, to exercise a conservative and constructively critical, highlighting its enormous potential, often uncritically emphasized for commercial reasons, but also its limits (and possible threats, such as the sci-fi dystopia of machines in power). This means, for example, detecting the lack of studies on the effectiveness of AI about important clinical outcomes, such as reducing morbidity/mortality or improving the quality of life of patients. The objectives should also include the level of satisfaction, both for physicians and patients, in the new relational context of integration of the digital world with the real one. It is ironic that just when the time in clinical practice is more and more limited, a deep reflection on the possible effects of the transformation underway, in terms of acceptance by physicians and all operators, of changes of professional role, of relationship with the patient, and essential training needs, is indispensable. The responsibility for the choice should remain personal, shared with the patient, both in the case that the professional decides to use the AI systems, and that he decides not to use them, even if in general it is more likely that a psychological mechanism of de-empowerment and delegation of intelligence to machines would develop: "I am not wrong," with the related medico-legal consequences to be considered in an area still substantially unknown. A sensitization of all healthcare workers is undoubtedly necessary to start a process of confrontation with citizens in order to establish strategies and policies towards a technology that, in the not distant future, is destined to change the essence of Medicine. As stated by Hegel "… Philosophy always comes too late. As thought in the world, it appears for the first time in time, after reality has completed its formation process and is beautiful and done... When philosophy paints in *chiaroscuro*, then an aspect of life has aged, and, since *chiaroscuro* does not allow itself to be rejuvenated, but only to recognize: the night of Minerva begins its flight in the twilight" [109]. The alternative is to lose the game or get to know the vastness of the phenomenon too late, when things are done, such as the night of Minerva, which comes when reality is "well done" [110, 111]. In confirmation of this context, we report four scientific publications that analyze cases where the results achieved from ML algorithms proved misleading, with the risk of supporting the physician in bad decisions.

The first case considered is described in an article that takes up the concept that ML techniques have already made extensive progress in many areas, including in Medicine, and will lead to important changes especially in the medical disciplines that require more accurate prognostic models such as oncology and the ones based on pathological recognition (e.g., radiology and pathology) [112]. But at the same time, this article also describes the so-called "controversial consequences" which may derive from the use of ML in clinical practice. In particular, the author deals with certain points:

1) deskilling: a consequence of the excessive dependence on the use of ML techniques to support clinical decisions in the long term could be the reduction of the competence level required to perform a medical function. If ML systems were wrong or temporarily stopped working, the phenomenon of deskilling would be more serious, as the excessive dependence on the use of technology would have led to a reduction of the physician's ability to analyze and make decisions. There are already real cases of this phenomenon, as shown below. In an analysis conducted by a group of researchers of the City University of London on the reading of 180 mammograms by 50 professionals, a decrease of 14.5% in diagnostic sensitivity has been documented for breast cancer detection in the most expert clinicians, when they were presented images of difficult reading accompanied by the interpretation by the computer. In contrast, only an increase of 1.6% in diagnostic sensitivity was detected with the computer support in the subset of less experienced clinicians when

cases of simpler interpretation were presented to them [113]. Another study on 30 patients showed a decrease in the diagnostic accuracy from 57% to 48% when electrocardiograms were noted with inaccurate computerized diagnoses [114]. These results highlight how the excessive trust on ML techniques lead to significantly influence the performance of the medical staff, but also that much research is still needed to precisely identify the characteristics of this phenomenon, especially the loss of "self-confidence" with the consequent inability to provide definitive interpretations and diagnoses;

2) data decontextualization: ML algorithms feed with large volumes of data, because only thanks to them they can process their predictions to support decisions. Relying on these systems requires that the data underlying the analyzes be considered as reliable and complete representations of the phenomena they should represent in a discreet form. This aspect can constitute a problem when the clinical context by its nature is not easily representable, unlike data that is easy to encode by numbers. The problem arises when ML systems have to manage context information difficultly representable by data, such as cultural, social, or psychological aspects of a patient, or organizational aspects of a hospital setting, which are necessary for a proper treatment, but that being not available to physicians could lead to incorrect decisions. An example in which the predictive ability of a ML algorithm as an aid to decisions has proven to be technically reliable, but potentially misleading, has been illustrated by Caruana et al. within a series of 14,199 patients with pneumonia, on whom different ML algorithms for the purpose of predicting the risk of mortality, and thus driving the management of patients in the intra- or extra-hospital setting, have been applied. The algorithms analyzed that patients suffering from asthma and pneumonia were at a lower risk of mortality compared to those who had pneumonia only, which may seem surprising. Clinicians, in fact excluded that asthma could be a protective factor for such patients attributing the error to a "confusing medical intervention." We should remember

that ML models do not apply rules to the provided data, but seek subtle relationships within it. The cause that made the algorithm produce these results, though correct, was due to the fact that in the hospitals where the study took place, patients with pneumonia and a history of asthma were admitted to intensive care unit and had a lower mortality, thanks to a better clinical check. Hence, the lack of the clinical context in the inclusion of ML algorithms led to suppose that patients suffering from pneumonia and asthma had better outcomes than those who had been diagnosed with pneumonia only, with a reduction in the mortality rate of approximately 50% (5.4% vs 11.3%, respectively). This shows how the lack of a variable in the data, i.e., the admission of the patient in intensive care unit, can induce algorithms, though perfect, to make a mistake because of the partiality of the data [115]. Prediction problems in biomedical sciences are generally quite difficult, partially due to incomplete knowledge of how the phenomenon of interest is influenced by the variables and measurements used for prediction, as well as a lack of consensus regarding the ideal predictors for specific problems. In these situations, a powerful approach to improving prediction performance is to construct ensembles that combine the outputs of many individual base predictors, which have been successful for many biomedical prediction tasks. Moreover, selecting a parsimonious ensemble can be of even greater value for biomedical sciences, where it is not only important to learn an accurate predictor, but also to interpret what novel knowledge it can provide about the target problem. Ensemble selection is a promising approach for this task because of its ability to select a collectively predictive subset, often a relatively small one, of all input base predictors. One of the most well-known algorithms for ensemble selection, Caruana et al.'s ensemble selection, generally performs well in practice, but faces several challenges due to the difficulty of choosing the right values of its various parameters. Since the choices made for these parameters are usually ad-hoc, good performance of the algorithm is difficult to guarantee for a variety of problems or datasets [116]. In general, it can be concluded that the lack of inclusion of clinical factors difficult to represent in ML systems could lead to contextual errors, and that therefore a high dependence on these systems could generate erroneous decisions with greater frequency and negative consequences; and

3) "black box" risk: ML algorithms are often referred to as "black box models," in the sense that the generated output is not easily understandable not only by medical users of these systems, but also by the engineers who developed them. In light of the high predictive accuracy of the current DL models, associated with their lack of transparency, and their inscrutability, we can speculate how these devices, should they become of common use, could considerably influence numerous aspects of the medical decision, even going to generate, in the long period, a sort of "oracular assignment," or excessive trust and therefore of potential dependence on these systems. This could change the way physicians learn, think, act, and interact with colleagues and patients. To relieve tension between accuracy and interpretability, efforts are being made to develop ML systems that independently provide explanations to physicians offering interactive tools to explore the implications of potential variables of exposure.

The second case considered is described in an article that shows that the search for online medical information has become increasingly frequent nowadays; in fact many people are looking for symptoms and medical remedies for injuries or health problems on the Internet. This behavior can be hazardous, as there is no guarantee regarding the quality of the information that can be found online, especially on unmoderated forums and social networks. The advent of these online tools has radically changed the scenario towards welfare structures; patients are faced with many more stimuli concerning those long supplied by healthcare professionals or by traditional support groups [117]. Despite the wide acceptance that online tools have received from healthcare community and patients, this exchange of information can lead to certain problems, such as the presence of incorrect, inaccurate, incomplete, improperly emphasized, ambiguous, or disputable medical advice. However, knowledge of others' experiences can support and positively reinforce the patient's confidence, confirm treatments, provide new alternatives when faced with problems, and help to alleviate loneliness by maintaining relationships with others. It is therefore essential to evaluate the quality of online information, both for those looking for certain topics for the first time, and to learn more about the understanding of disease within specific communities. An interesting phenomenon that occurs is "prosumerism," as the online sharing of information, stimulated by anonymity, can increase frankness and sincerity when personal problems and experiences are touched. This may be true for both sporadic and chronic patients; the former seek information about unknown symptoms, while the latter seek updates and new solutions to their known problems. Chronic patients spend a good part of their life optimizing the treatment and their lifestyle, to achieve a higher quality of life, acquiring a deep "know-how" and developing a high degree of self-awareness on how diseases should be managed concerning the occasional patient. Sharing one's experiences online could be useful for both new patients, and also for medical researchers: as the former could learn to handle new situations, while the latter could acquire a more in-depth knowledge about his clinical area of study. This article focuses on the role of a popular online social networking tool, like Facebook, for a particular class of chronically ill people, Crohn's disease (CD) patients. This disease can be treated, but not definitively cured; typically, people who suffer from it fight for their whole life with their symptoms, potentially sharing information and experiences online. An analysis focused on the online debate on CD has been performed, where patients' comments about specific topics, such as possible symptomatic causes, symptoms, treatments, and side effects were studied. This analysis led to two main results: the information on CD is easier to find on the Facebook page than on the Twitter one, and the pharmaceutical treatment that generates the most significant number of posts is iniximab. The analysis was performed in three different steps:

1) through data mining methodologies, analyzing the feeling that emerges at the discussion of a specific topic;
2) by understanding how the online discussion of a certain treatment influences the mood, which can be detected by post. In this second step, iniximab has been identified as the treatment that most influences the patients' mood; and
3) by reading and interpreting one by one the posts concerning iniximab, looking for one or more rules that would characterize behavior, perception, and online published comments. This information has been related to significant scientific literature in relation to iniximab use.

To detect posts where people discuss about CD on Facebook, the public pages dedicated to this disease have been identified through a quantitative method. The analysis then developed with the following integrity check: a classification of the 20 most prolific authors, reading the content of their posts. The result of this analysis is that most of the authors on Facebook are patients, as they usually share personal experiences related to their therapeutic path, contrary to what happens on other social networks, such as Twitter. The problem was finding a reliable way to evaluate the user's feeling relating to a post in which the drug was mentioned. To do that, OpinionFinder was used, a system that can process a body of text and identification sentences, returning a value that satisfies the feeling of each sentence, classifying it as neutral, positive, or negative. So, OpinionFinder was used with the aim of measuring how a feeling expressed by a patient may change in relation to the administration of a specific drug. Indeed, this work's relevance is not finding an absolute value of feeling, but the variation of mood expressed by patients in the various posts. At this point, the second step was built to check how the discussion of a specific treatment could influence/cause positive or negative moods, starting from an analysis on Facebook posts. The result of this analysis, also using Granger's analysis, identified that iniximab influences mood in a positive or negative way, without being able to decide for one or the other. For this reason, the work has been extended with a qualitative analysis to analyze the posts through an inductive approach; this technique was used to analyze the data collected by social media for the purpose of identifying common/uncommon themes to traditional medical research. The exciting result of the study was to note that it is possible to find common points and identify norms that emerge from both contexts. In detail, the study starts with a quantitative analysis on pharmacological treatments for CD patients through two questions:

1) how do people consider a treatment being positive or negative by their online posts? and
2) how is the mood expressed by patients online influenced by the use of certain drugs?

All drugs with the same active principle were taken into consideration in the analysis. To view the sentimental evolution of a certain treatment with a graphic, the sentiment value was calculated as the difference between the positive and negative sentiments expressed during a week. For example, if OpinionFinder finds 10 positives and 4 negatives, the value of sentiment for that week will be 6. The results led to the decision to examine only three drugs (adalimumab, azathioprine, and iniximab) although this analysis did not lead to an exhaustive understanding of the mood expressed in the posts. An assessment was then made of the opinions expressed during the discussions on a certain treatment, e.g., by checking if the discussions influence/cause a change of mood on Facebook pages about CD. In the search for a relationship between a discussion on a pharmacologic treatment and positive/negative mood, Granger analysis was used and only iniximab demonstrated a significant statistical correlation. Concluding, iniximab is able to influence the community of people affected by CD, but we are not sure whether in a positive or negative way. The controversial result of the quantitative analysis of Facebook posts brings here to a further investigation based on how iniximab really influences the mood of the community affected by CD. A total of 241 Facebook posts were analyzed, of which 115 were discarded because they were considered neutral. This analysis sought, where possible, to classify the final mood of each post. A Likert scale between -2 and 2 was used, where the values represented:

1) very unsatisfied (-2);
2) not satisfied (-1);
3) uncertain (0);
4) satisfied (1); and
5) very satisfied (2).

It was not possible to determine whether iniximab use causes a positive or negative feeling, as the degree of satisfaction was equally distributed. Subsequently, an analysis was carried out also taking into account the duration of treatment; it was possible to extract, in addition to a value that represents the mood of the post, also another numeric value that represents the duration of treatment, to look for a correlation between the duration of treatment and patients' mood. Surprisingly, the data looks much more interesting grouped based on the different treatment durations, noting two results:

1) almost 30% of patients express a degree of unsatisfaction concerning iniximab use; and
2) a longer duration of treatment corresponds to a higher level of satisfaction.

Analyzing the duration of treatments on a case by case basis, it has been shown that:

1) a one month-long treatment led to unsatisfaction towards iniximab;
2) in the case of a treatment lasting between 2 months and two years, more positive posts than negative ones were counted, with a slight increase in the trend of satisfaction;
3) in the case of a treatment lasting between 2 and 5 years, the result is unexpectedly 50% – 50%; and
4) in the case of a treatment lasting between 5 and 13 years, patients are clearly satisfied.

A review of medical literature was conducted to check whether the described results could also be significant from a medical point of view. A first emerged data is the direct correlation between the small number of patients involved in the clinical study and the large number of positive results. One possible explanation could be the increased ability of physicians to check the evolution of the disease and the greater adequacy of a certain treatment for each individual patient. On the contrary, when the number of patients involved in a study is higher, also unsatisfaction with the drug increases. The conclusion of the work reports controversial results regarding CD patients taking iniximab; such results can be understood when the duration of treatment and the size of the population involved in the study are taken into consideration. Indeed, a parallelism between the two data sources (Facebook posts and clinical studies described in the updated medical literature) exists, e.g., as the percentage of the analyzed population grows, also the percentage of patients that report a negative experience increases. A possible development of this work would be to extend necessary information on patients such as age, sex, and other diseases, which would allow a more effective analysis; unfortunately, today social media do not allow any kind of deduction that can be considered scientifically valid. So, to date the analysis of a social media network like Facebook cannot support the medical community with additional information that conversely would be very useful for herself to facilitate its own education.

The third case considered is described in an article that reports how much the forecast accuracy is fundamental in the correct execution of ML algorithms [118]. The customarily used approach to verify this accuracy is cross-validation (CV). Two popular CV methods are subject-wise and record-wise. In this technique, the sample set is divided into *K* partitions of the same size. At each iteration of the method, one part of the partition is used for testing and the rest for training, with which the model is built. The purpose is to evaluate the ability of the algorithm to generalize new data. For CV to be valid, the training and testing sets must be independent. The concept of independence depends on the scenario of use:

1) diagnosis: global models are developed that can be used for new subjects, the used CV method must be subject-wise in the sense that the training and the testing set must contain records of different subjects; and
2) prognosis: personal models are needed that can predict future clinical states of a particular subject, dividing the data according to time, so that the training and testing set contain records of the same subject in different periods, therefore the used method is record-wise.

Therefore, the choice of CV method depends on the use case. In a diagnosis scenario, where we want to build a model which will then be used to classify new subjects, it is necessary to use the subject-wise CV. Nevertheless, the record-wise method can also be used, which randomly subdivides the data in training and testing sets regardless of the subjects to which they belong. So, the records of the same subject are found in both the training and testing phase. In this way, the algorithm can find an association between the unique characteristics of an individual (such as the walking speed) and its clinical condition, which automatically improves prediction accuracy. The record-wise CV method may lead to an overestimation of the accuracy of algorithm prediction. The article gives an example that clarifies why the use of the CV record-wise method in a diagnosis is misleading [119]. The analyzed case considers four individuals, two healthy and two affected by PD; the ML algorithm estimates whether a person is affected by PD based on his walking speed. Healthy subjects have an average speed of 1 m/s and 0.4 m/s and the affected ones 0.6 m/s and 0.2 m/s, respectively. Using the subject-wise CV method, we will not be able to predict the performances of the healthy subject or of the affected one, obtaining a prediction accuracy of about 50%. If, on the other hand, we use a record-wise CV, we have 10 records available for each subject, 9 measurements are used to predict the tenth. We will therefore be able to know the condition of the subject. By this method, the identification of the individual would be relatively easy, and would replace the recognition of the disease, which we know to be instead complex. In this case, the record-wise CV method would give a 100% accuracy, which is not supported by the data, and therefore the algorithm will not generalize. Another check is carried out in the article to confirm the difference between the two presented methods of record-wise and subject-wise CV. The employed dataset takes into consideration a set of recognition of human activities, publicly available, which contains recordings of 30 subjects performing 6 activities: sitting, standing, walking, going up and going down the stairs, and lying down. The used data were collected using recordings by the accelerometer and the gyroscope sensors of the smartphone. In the article, RF, an ensemble of DT, where each tree provides a prediction on the input data class, has been utilized as classifier. The end result of the RF is determined by aggregating all the predictions of the individual trees. Each individual tree in a RF refers only to a subset of characteristics and samples of input data. An RF has, therefore, fewer parameters to optimize, making this method less prone to overfitting and an optimal way to classify new data. For both CV methods, 2, 10, or 30 subjects and k-fold CV with $k = 2$, 10, or 30 were used. As for subject-wise, the data was divided in such a way that the training and testing phase would have contained records of different subjects. In both methods, the classifier was trained on all records except one, on which the testing phase was conducted. Initially, the performance of the CV subject-wise method was assessed. At the time when 2 folds and 2 subjects were used, the error rate of the classification algorithm was initially detected at 27% but, increasing the number of subjects taken into consideration to a maximum of 30, it has decreased to 9%. By increasing the number of folds to a maximum of 30, data from multiple subjects trained the classifier and the error rate reached 7%. Subsequently, the second method (record-wise) was used and the same procedure has been performed to analyze how the error varied by number of folds and subjects. It is interesting to observe how the starting classification error, even when data of only two subjects and folds have been taken into consideration, was already at 2%, and increasing the number of these (subjects and folds) the error has not changed significantly. It should be emphasized that, using this method, the precision of classification on this dataset has been overestimated. The central theme of the article is to assess the reliability of ML accuracy and wearable sensor technology to predict clinical results. Taking into consideration the dataset, it has been verified as the CV record-wise method produces high but misleading accuracy. Only one way in which CV can be misused in ML was considered in the article. In many prediction algorithms, in addition to the main parameters, also hyper-parameters must be considered, which are very often chosen in such a way to minimize the prediction error on the data set. The correct way to adjust these hyper-parameters is further divide the training set into subsets of training and validation sets and minimize the prediction error in these subsets compared to doing it over the entire data set. This approach allows a correct evaluation of the generalizability of the algorithm. While only one dataset was used in the article, it is possible to use any clinical data set to show how, mixing among subjects in the training and validation set phase, it is possible to increase, in an artificial way, prevision accuracy. As in Medicine clinical records often include information on physical or physiological characteristics of an individual, such as the waistline or the blood group, which don't change much over time, if we use the record-wise method, the algorithms will already know the clinical outcome for an individual with specific characteristics. On the other hand, the other method will ensure that there are no shortcuts to the algorithm. Therefore, the choice of the CV method is particularly important in clinical prevision applications. Since ML algorithms are increasingly used, it is necessary to evaluate through CV their accuracy since inadequate validation procedures could lead to unreliable results, which would lead to the problem of irreproducibility of the research results and therefore undermine trust in Medicine and data science.

The fourth case is taken from a published article that highlights the fundamental role played by big data in the diffusion of ML algorithms to support clinical decisions in Medicine. A simple and inexpensive way to extract as much more accurate medical information and improve monitoring, diagnostics, and healthcare treatment customization is the use of devices like smartphones and smartwatches [120]. The article focuses on mental well-being as it is a pathology that requires a long-term monitoring which is therefore made sustainable thanks to commonly used devices and sensors. Typical algorithms related to mental well-being predictions apparently seem to work well, but in reality, only because they are guessing the average personal states of individuals. This example highlights how easy it is to get falsely optimistic results when the result obtained by the algorithm is compared with the baseline of the population than that of the individual. This approach is followed by most studies (approximately 77%) in literature, the baseline of the population is used as a comparison method. The article proposes a new measure, the "user lift," which measures the benefit of the algorithm relating to the one-person model to avoid to obtain falsely optimistic conclusions. To evaluate the accuracy of the algorithm in predicting a binary result, like for example a happy day compared to a sad one, the prediction error is considered, which is the percentage of observations that have been wrongly predicted.

On the contrary, to evaluate the accuracy of an algorithm for predicting an individual's level of happiness or stress, the *root-mean-square error* is used, which indicates the mean quadratic discrepancy between observed and estimated data values. In both cases, errors with low values indicate that the algorithm is well predicting the status of an individual, instead higher values indicate a more significant prediction error. The two baselines that we consider are the personal one, that is of an individual, and the population one. The personal one supposes that each individual is always in a constant state, but can differ from person to person, instead, the population one requires every individual is in the same state. The "user lift" method is proposed to evaluate whether the algorithm is doing better predictions than just guessing the state of the individual. It is the improvement of the algorithm predictions with respect to the personal baseline. The article presents an example of how the use of ML techniques can lead to falsely optimistic conclusions, particularly in the prediction of an individual's mental state of well-being. Two datasets were considered, both with data collected from the smartphones of the analyzed subjects, in particular, their global *positioning* system (GPS) position and their stress and happiness levels:

1) StudentsLife Dataset: relating to a group of students from an American university, daily stress measures are collected on a five-point Likert scale [121]; and
2) MIT Friends and Family dataset: related to a group of members from another American university and their relatives. Here we consider the nine-point Likert happiness scale and the Seven-point Likert stress scale [122].

To derive significant localization and mobility features, the data have not been processed. The first method uses a Gaussian mixture model at all the localization samples for each participant to identify their positions. Participants were assumed to attend at most twenty locations, the main ones being home and the workplace that have been considered as complete clustering. The second method uses K-means clustering for fixed locations only. StudentLife dataset included a prediction that the individual was stationary in one point or moving, but not the other dataset. We consider this second set as stationary clustering and the night cluster as the cluster where each individual spends more time between 12 PM – 6 AM. Downstream of the identification of the datasets and the execution of methods of data preparation, two prevision activities have been carried out:

1) predicting if an individual was happy/stressed or less on a given day; and
2) predicting the average level of happiness or stress that a participant has reported on a particular day.

To assess stress or happiness levels on a given day, the averages of all the answers that a participant reported on that day were calculated on the Likert scale. To find out if the participant was happy (or stressed) or less, a threshold has been defined on the daily average on value to distinguish when students reported a stressful situation or not. For StudentLife dataset, the threshold is "a little stressed" for Friends and Family the average value of the Likert scale is used as the threshold. The results obtained from this example show how baselines are important in ML in Medicine and how they are used in practice. The personal model produces a minor error concerning that of the population and the baseline of the population, but not less than the personal baseline, so the algorithms simply predict the most frequently reported states. To avoid comparison with the wrong baseline, it is proposed to use statistical tests with the user lift metric to prove that an algorithm is doing much better than the personal baseline. The user lift technique calculates the difference of the personal model with the personal baseline.

Positive user lifts indicate that a model is better than the personal baseline, that the algorithm predictions are more accurate than one supposes provided an individual is in his average state. The user's user lifts are not significantly higher than zero, and therefore the models do not provide better results compared to constant personal baselines. In this study we have shown, with examples of stress and happiness using two public datasets, how easy it is that ML algorithms are promising when compared to the wrong baselines. Individuals surprisingly report small changes in their mental state, so predicting that an individual is in his most frequent state of mind is correct in the majority of cases. As a result, when an algorithm is compared with a population baseline that always predicts that all people are in the same state, the results of the algorithm may seem accurate even if not ameliorative in predicting that each individual is in his most frequent state. Population baseline was found to be used in 77% of publications, and when the person one is used, the result doesn't change much, rather sometimes it gets worse. A limitation on the chosen dataset was that the study groups were not clinical populations, the sample size was small, and the duration of the study was limited. While the target population usually taken into consideration should be individuals with mood disorders, study groups taken into consideration are often small and mentally healthy. People with mood disorders may report more status changes compared to the normal population. Greater variability would reduce the probability of falsely optimistic results. The ability to predict significant personal signals for medical monitoring, like mental well-being, could significantly improve personalized care allowing just-in-time approaches and personalized interventions. However, some pitfalls have been highlighted in the evaluation of the algorithms for this application that can lead to falsely optimistic results by providing unfounded optimism. To limit them, it has been proposed to use the user lift metric, which is directly concentrated on the individual.

CONCLUSION

Digital transformation of society has been characterizing, for years now, the life of individuals and their professional activities. In particular, the most innovative aspects of this transformation are represented by the progressive introduction of AI into our daily life and how much ML algorithms can help us simplify it. These technologies have now established themselves in all business sectors, both production and services, and among these, the medical sector stands out, where there is great hope in the contribution they will be able to give in the next years to healthcare and more generally to individual quality of life. The introduction of AI and ML algorithms in Medicine has allowed us to achieve positive results already and to foresee the promising future prospects. An intelligent machine is based on technologies that, thanks to the development of specific algorithms, make these machines "intelligent," that is capable of developing a "human" activity without the support of a man. But there are also negative aspects and concerns related to the introduction of these systems as decision-making support to medical caregivers in the performance of their activities. There are currently criticisms and perplexities related to AI and ML in Medicine, or on the possibility of not having a real and precise decision-making support in healthcare, but on the contrary of obtaining unexpected or even misleading results.

Analyzing the available literature, four articles were identified which describe concrete cases that prove how specific ML algorithms may lead to incorrect assessments and, more generally, to critical issues on medical caregivers. In the first case, the difficulty of giving a complete computerized training to the clinical context, which, by its very nature, is highly articulated. This lack within ML algorithms may adversely affect their functioning and therefore produce incorrect and even misleading results in the physician's decision-making process. Another critical element highlighted is the issue of over-reliance; that is how the introduction of decision support of ML systems can bring over time to an excessive dependence by physicians on these technologies, and therefore as main consequence the deskilling of the healthcare staff. In the second case, it was evaluated as the analysis of the use of social networks, for example Facebook, by the community of CD patients may be of assistance to medical activity. The conclusion we have reached is that to date, social networks cannot allow scientifically valid deductions to understand what is happening to a patient, in particular, due to the incompleteness of the information. In the third case, the importance of checking the accuracy of ML algorithms was highlighted, which can be obtained thanks to CV. We have also proved how the choice of the CV method must be strictly linked to the use scenario, so as not to lead to an overestimation of the algorithm accuracy. In the fourth case, it was highlighted that the choice of the correct dataset is fundamental in order not to produce misleading results; in particular we have demonstrated how the model can result in inaccurate if we compare a single individual with the baseline of the entire population [123].

The diversity and huge omics data take biology and biomedicine research and application into a significant data era, just like that prevalent in human society a decade ago. They are opening a new challenge from horizontal data ensemble (e.g., the similar types of data collected from different labs or companies) to vertical data ensemble (e.g., the different types of data collected for a group of person with match information), which requires the integrative analysis in biology and biomedicine and also asks for emergent development of data integration to address the great changes from previous population-guided to newly individual-guided investigations. Data integration is a useful concept to solve complex problem or understand the complicate system. Several benchmark studies have revealed the heterogeneity and trade-off that existed in the analysis of omics data. Integrative analysis can combine and investigate many datasets in a cost-effective reproducible way. Current integration approaches on biological data have two modes: one is "bottom-up integration" mode with manual follow-up integration, and the other one is "top-down integration" mode with follow-up in silico integration [124]. AI is a very active computer science research field aiming to develop systems that mimic human intelligence and is helpful in many human activities, including Medicine. Some examples of the exploiting of AI techniques, in particular automatic classifiers such as ANN, SVM, classification tree, and ensemble methods like RF, are the ability to analyze findings obtained by positron emission tomography (PET) or single-photon emission tomography (SPECT) scans of patients with neurodegenerative diseases, in particular AD, or to preprocess data and reduce their dimensionality via feature selection or projection in a more representative domain (Principal Component Analysis [PCA] or Partial Least Squares [PLS] are examples of such methods); this is a crucial step while dealing with medical data, since it is necessary to compress patient information and retain only the most useful in order to discriminate subjects into normal and pathological classes. The increasing development of CAD systems is very promising to contribute to the diagnostic process [125].

Despite the accelerating pace of scientific discovery, current clinical research enterprise does not sufficiently address pressing clinical questions. Given the constraints on clinical trials, for a majority of clinical questions, the only relevant data available to aid in decision-making are based on observation and experience. Randomized controlled trials (RCT) are not at risk for extinction, but innovations in statistics, ML, and big data analytics may generate a completely new ecosystem for exploration and validation. However, it is necessary to underline how these technologies can give a considerable contribution to medical practice (e.g., by processing a lot of data in a concise time, an activity impossible for the human brain), but at the same time, they cannot ignore the human contribution by the physician, who must always give a final sense to the analyzed data also by virtue of the non-deterministic nature of medical science [126]. The human mind learns to make complex associations and to generalize them from a limited amount of data. At the same time, "machines" require many more examples, besides not being provided with common sense.

On the other hand, however, they can process and learn from massive quantities of data, quantities that our brain would not even be able to imagine, and without incurring human errors due to carelessness or fatigue.

Having said that, the physician-patient relationship is considered essential and the evolution of Medicine towards digital can only improve and enrich the patient's diagnosis, treatment, and care process, reducing human errors, increasing the time the physician dedicates to the patient, lowering mortality rates and healthcare costs. But it cannot totally replace the human being [127]. The human mind is unable to store such a large amount of information, such as clinical reports and previous diagnoses, and this is why technology comes to our rescue. The implementation of ML is therefore an opportunity to be exploited in the health sector. It is right, however, to make a clarification: every AI tool is a support to be used and it is unthinkable to rely only on it. Medicine and technology, as we said, must "collaborate" and must not be mutually exclusive [128]. It is reasonable to think, in the hospital of tomorrow (or, in some cases, today), that the clinician, before deciding (because he/she will always do it), asks for a second opinion from an AI. With benefits for the whole society (lower costs, greater effectiveness). This is an immeasurably more likely future than the one revealed in the widespread dystopian visions in which AI takes over. But on these you can write a book and make a film. On CAD that helps doctors, we are afraid not [129].

REFERENCES

[1] Minini, A. (2020). *Ensemble learning (machine learning technique)*. Andrea Minini. Online Personal Knowledge Base/Beta, January 28, 2020. http://www.andreaminini.com/ai/machine-learning/ensemble-learning.

[2] Baskin, I. I., Winkler, D., and Tetko, I. V. (2016). A renaissance of neural networks in drug discovery. *Expert Opin Drug Discov*, 11 (8): 785-795.

[3] Kim, J., Ahn, J., and Cho, S. (1995). Ensemble competitive learning neural networks with reduced input dimension. *Int J Neural Syst*, 6 (2): 133-142.

[4] Soto, V., García-Moratilla, S., Martínez-Muñoz, G., Hernández-Lobato, D., and Suárez, A. (2014). A double pruning scheme for boosting ensembles. *IEEE Trans Cybern*, 44 (12): 2682-2695.

[5] Turing, A. M. (1937). On Computable Numbers, with an Application to the Entscheidungsproblem. *Proceedings of the London Mathematical Society*, s2-42 (1): 230-265.

[6] McCulloch, W. S., and Pitts, W. (1943). A logical calculus of the ideas immanent in nervous activity. *Bulletin of Mathematical Biophysics*, 5: 115-133.

[7] Edmonds, D. S. Jr. (1950). Recollections of my Wartime and University Experiences. *Engineering and Technology History Wiki*, Updated January 13, 2015. https://ethw.org/First-Hand:Recollections_of_my_Wartime_and_University_Experiences_in_Nuclear_Physics.

[8] Turing, A. M. (1950). Computing machinery and intelligence. *Mind*, 59: 433-460.

[9] Samuel, A. L. (1953). Computing bit by bit, or Digital computers made easy. *Proceedings of the Institute of Radio Engineers*, 41: 1223-1230.

[10] McCarthy, J., Minsky, M. L., Rochester, N., and Shannon, C. E. (2006). A Proposal for the Dartmouth Summer Research Project on Artificial Intelligence, August 31, 1955. *AI Magazine*, 27 (4): 12.

[11] Rosenblatt, F. (1957). *The Perceptron – A perceiving and recognizing automaton. Report 85–460–1*. Buffalo: Cornell Aeronautical Laboratory.

[12] Khudorozhkov, R., Emelyanov, K., Koryagin, A., and Kozhevin, A. (2017). RadIO library for data science research of CT images. 2017. CERN Data Centre & Invenio, January 21, 2020. https://zenodo.org/record/1156364#.XoeaRUBuLIV.

[13] Lagasse, J. (2017). Art Medical's smart tubes to collect patient data, reduce complications. *Mobihealthnews*, February 7, 2020. https://www.mobihealthnews.com/content/art-medicals-smart-tubes-collect-patient-data-reduce-complications.

[14] Samuel, A. L. (1959). Some Studies in Machine Learning Using the Game of Checkers. *IBM Journal of Research and Development*, 44 (1.2).

[15] Anwar, H., Qamar, U., and Muzaffar Qureshi, A. W. (2014). Global optimization ensemble model for classification methods. *Scientific World Journal*, 2014: 313164.

[16] Blower, P. E., and Cross, K. P. (2006). Decision tree methods in pharmaceutical research. *Curr Top Med Chem*, 6 (1): 31-39.

[17] Che, D., Liu, Q., Rasheed, K., and Tao, X. (2011). Decision tree and ensemble learning algorithms with their applications in bioinformatics. *Adv Exp Med Biol*, 696: 191-199.

[18] Zhou, Y., and McArdle, J. J. (2015). Rationale and Applications of Survival Tree and Survival Ensemble Methods. *Psychometrika*, 80 (3): 811-833.

[19] Collell, G., Prelec, D., and Patil, K. R. (2018). A simple plug-in bagging ensemble based on threshold-moving for classifying binary and multiclass imbalanced data. *Neurocomputing*, 275: 330-340.

[20] Candelieri, A., and Conforti, D. (2010). A Hyper-Solution Framework for SVM Classification: Application for Predicting Destabilizations in Chronic Heart Failure Patients. *Open Med Inform J*, 4: 136-140.

[21] Alhusain, L., and Hafez, A. M. (2017). Cluster ensemble based on Random Forests for genetic data. *Bio Data Min*, 10: 37.

[22] Sloutsky, R., Jimenez, N., Swamidass, S. J., and Naegle, K. M. (2013). Accounting for noise when clustering biological data. *Brief Bioinform*, 14 (4): 423-436.

[23] Hebb, D. O. (1949). *The Organization of Behavior*. New York: Wiley & Sons.

[24] Hart, E., and Sim, K. (2018). On Constructing Ensembles for Combinatorial Optimisation. *Evol Comput*, 26 (1): 67-87.

[25] Zorman, M., and Verlic, M. (2009). Explanatory approach for evaluation of machine learning-induced knowledge. *J Int Med Res*, 37 (5): 1543-1551.

[26] de Kerckhove, D., and Buffardi, A. (2011). *Digital knowledge*. Naples: Liguori Editore.

[27] (2020). Big data in medicine: Development with machine learning. *Ianiri Informatics*, February 7, 2020. https://www.gipo.it/blog/medicina-innovazione/big-data-medicina-2/.

[28] D'Avenio, G., and Grigioni, M. (2018). Deep learning and healthcare, all the unknowns of the AI "black box." *ICT & Strategy*, May 14, 2019. https://www.agendadigitale.eu/cultura-digitale/deep-learning-e-sanita-tutte-le-incognite-della-scatola-nera-della-ai/.

[29] Wall, R., Cunningham, P., Walsh, P., and Byrne, S. (2003). Explaining the output of ensembles in medical decision support on a case by case basis. *Artif Intell Med*, 28 (2): 191-206.

[30] Kellner, T. (2015). Dr. Data: How the Health Cloud Will Help Doctors Combat Disease. GE Reports, November 29, 2015. *General Electric*, March 7, 2020. https://www.ge.com/reports/dr-data-how-the-health-cloud-will-help-doctors-combat-disease/.

[31] Onofrey, J. A., Staib, L. H., Huang, X., Zhang, F., Papademetris, X., Metaxas, D., Rueckert, D., and Duncan, J. S. (2020). Sparse Data-Driven Learning for Effective and Efficient Biomedical Image Segmentation. *Annu Rev Biomed Eng*, March 13, 2020. [Epub ahead of print].

[32] Gulshan, V., Peng, L., Coram, M., Stumpe, M. C., Wu, D., Narayanaswamy, A., Venugopalan, S., Widner, K., Madams, T., Cuadros, J., Kim, R., Raman, R., Nelson, P. C., Mega, J. L, and Webster, D. R. (2016). Development and Validation of a Deep Learning Algorithm for Detection of Diabetic Retinopathy in Retinal Fundus Photographs. *JAMA*, 316 (22): 2402-2410.

[33] Esteva, A., Kuprel, B., Novoa, R. A., Ko, J., Swetter, S. M., Blau, H. M., and Thrun, S. (2017). Dermatologist-level classification of skin cancer with deep neural networks. *Nature*, 542 (7639): 115-118.

[34] Ghafoorian, M., Karssemeijer, N., Heskes, T., van Uden, I. W. M., Sanchez, C. I., Litjens, G., de Leeuw, F. E., van Ginneken, B., Marchiori, E., Platel, B. (2017). Location Sensitive Deep

Convolutional Neural Networks for Segmentation of White Matter Hyperintensities. *Sci Rep*, 7 (1): 5110.

[35] Rajpurkar, P., Irvin, J., Zhu, K., Yang, B., Mehta, H., Duan, T., Ding, D., Bagul, A., Ball, R. L., Langlotz, C., Shpanskaya, K., Lungren, M. P., and Ng, A. Y. (2017). *CheXNet: Radiologist-Level Pneumonia Detection on Chest X-Rays with Deep Learning*. Buffalo: Cornell University.

[36] Sinclair, M., Baumgartner, C. F., Matthew, J., Bai, W., Martinez, J. C., Li, Y., Smith, S., Knight, C. L., Kainz, B., Hajnal, J., King, A. P., and Rueckert, D. (2018). Human-level Performance on Automatic Head Biometrics in Fetal Ultrasound Using Fully Convolutional Neural Networks. *Conf Proc IEEE Eng Med Biol Soc*, 2018: 714-717.

[37] Ziegler, A., DeStefano, A. L., König, I. R., Bardel, C., Brinza, D., Bull, S., Cai, Z., Glaser, B., Jiang, W., Lee, K. E., Li, C. X., Li, J., Li, X., Majoram, P., Meng, Y., Nicodemus, K. K., Platt, A., Schwarz, D. F., Shi, W., Shugart, Y. Y., Stassen, H. H., Sun, Y. V., Won, S., Wang, W., Wahba, G., Zagaar, U. A., and Zhao, Z. (2007). Data mining, neural nets, trees – Problems 2 and 3 of Genetic Analysis Workshop 15. *Genet Epidemiol*, 31 (Suppl 1): S51-60.

[38] Manilich, E. A., Özsoyoğlu, Z. M., Trubachev, V., and Radivoyevitch, T. (2011). Classification of large microarray datasets using fast random forest construction. *J Bioinform Comput Biol*, 9 (2): 251-267.

[39] Wang, Y., Goh, W., Wong, L., and Montana, G.; Alzheimer's Disease Neuroimaging Initiative. (2013). Random forests on Hadoop for genome-wide association studies of multivariate neuroimaging phenotypes. *BMC Bioinformatics*, 14 (Suppl 16): S6.

[40] Song, L., Langfelder, P., and Horvath, S. (2013). Random generalized linear model: a highly accurate and interpretable ensemble predictor. BMC Bioinformatics, 14: 5.

[41] Tan, A. C., and Gilbert, D. (2003). Ensemble machine learning on gene expression data for cancer classification. *Appl Bioinformatics*, 2 (3 Suppl): S75-S83.

[42] Dudoit, S., and Fridlyand, J. (2003). Bagging to improve the accuracy of a clustering procedure. *Bioinformatics*, 19 (9): 1090-1099.

[43] Midorikawa, Y., Tsuji, S., Takayama, T., and Aburatani, H. (2012). Genomic approach towards personalized anticancer drug therapy. *Pharmacogenomics*, 13 (2): 191-199.

[44] Mohammed, A., and Guda, C. (2011). Computational Approaches for Automated Classification of Enzyme Sequences. *J Proteomics Bioinform*, 4: 147-152.

[45] Korvigo, I., Afanasyev, A., Romashchenko, N., and Skoblov, M. (2018). Generalising better: Applying deep learning to integrate deleteriousness prediction scores for whole-exome SNV studies. *PLoS One*, 13 (3): e0192829.

[46] Rajaraman, S., Sornapudi, S., Kohli, M., and Antani, S. (2019). Assessment of an ensemble of machine learning models toward abnormality detection in chest radiographs. *Conf Proc IEEE Eng Med Biol Soc*, 2019: 3689-3692.

[47] Rajaraman, S., Candemir, S., Xue, Z., Alderson, P. O., Kohli, M., Abuya, J., Thoma, G. R., and Antani, S. (2018). A novel stacked generalization of models for improved TB detection in chest radiographs. *Conf Proc IEEE Eng Med Biol Soc*, 2018: 718-721.

[48] Izadyyazdanabadi, M., Belykh, E., Mooney, M. A., Eschbacher, J. M., Nakaji, P., Yang, Y., and Preul, M. C. (2018). Prospects for Theranostics in Neurosurgical Imaging: Empowering Confocal Laser Endomicroscopy Diagnostics via Deep Learning. *Front Oncol*, 8: 240.

[49] Castillo-Barnes, D., Ramírez, J., Segovia, F., Martínez-Murcia, F. J., Salas-Gonzalez, D., and Górriz, J. M. (2018). Robust Ensemble Classification Methodology for I123-Ioflupane SPECT Images and Multiple Heterogeneous Biomarkers in the Diagnosis of Parkinson's Disease. *Front Neuroinform*, 12: 53.

[50] Islam, J., and Zhang, Y. (2018). Brain MRI analysis for Alzheimer's disease diagnosis using an ensemble system of deep convolutional neural networks. *Brain Inform*, 5 (2): 2.

[51] Wiemer, J. C., and Prokudin, A. (2004). Bioinformatics in proteomics: application, terminology, and pitfalls. *Pathol Res Pract*, 200 (2): 173-178.

[52] Richardson, A. M., and Lidbury, B. A. (2013). Infection status outcome, machine learning method and virus type interact to affect the optimised prediction of hepatitis virus immunoassay results from routine pathology laboratory assays in unbalanced data. *BMC Bioinformatics*, 14: 206.

[53] Datta, S. (2013). Feature selection and machine learning with mass spectrometry data. *Methods Mol Biol*, 1007: 237-262.

[54] Ezzat, A., Wu, M., Li, X. L., and Kwoh, C. K. (2016). Drug-target interaction prediction via class imbalance-aware ensemble learning. *BMC Bioinformatics*, 17 (Suppl 19): 509.

[55] Tan, K. K., Le, N. Q. K., Yeh, H. Y., and Chua, M. C. H. (2019). Ensemble of Deep Recurrent Neural Networks for Identifying Enhancers via Dinucleotide Physicochemical Properties. *Cells*, 8 (7).

[56] Nguyen, Q. H., Nguyen-Vo, T. H., Le, N. Q. K., Do, T. T. T., Rahardja, S., and Nguyen, B. P. (2019). iEnhancer-ECNN: identifying enhancers and their strength using ensembles of convolutional neural networks. *BMC Genomics*, 20 (Suppl 9): 951.

[57] Do, D. T., and Le, N. Q. K. (2020). Using extreme gradient boosting to identify origin of replication in Saccharomyces cerevisiae via hybrid features. *Genomics*, 112 (3): 2445-2451.

[58] Diao, R., Chao, F., Peng, T., Snooke, N., and Shen, Q. (2014). Feature selection inspired classifier ensemble reduction. *IEEE Trans Cybern*, 44 (8): 1259-1268.

[59] Bouktif, S., Hanna, E. M., Zaki, N., and Abu Khousa, E. (2014). Ant colony optimization algorithm for interpretable Bayesian classifiers combination: application to medical predictions. *PLoS One*, 9 (2): e86456.

[60] Burai, P., Hajdu, A., Manuel, F. E., and Harangi, B. (2018). Segmentation of the uterine wall by an ensemble of fully convolutional neural networks. *Conf Proc IEEE Eng Med Biol Soc*, 2018: 49-52.

[61] Siristatidis, C., Pouliakis, A., Chrelias, C., and Kassanos, D. (2011). Artificial intelligence in IVF: a need. *Syst Biol Reprod Med*, 57 (4): 179-185.

[62] Gürgen, F. S., Sihmanoglu, M., and Varol, F. G. (1995). The assessment of LH surge for predicting ovulation time using clinical, hormonal, and ultrasonic indices in infertile women with an ensemble of neural networks. *Comput Biol Med*, 25 (4): 405-413.

[63] Ilhan, H. O., Sigirci, I. O., Serbes, G., and Aydin, N. (2020). A fully automated hybrid human sperm detection and classification system based on mobile-net and the performance comparison with conventional methods. *Med Biol Eng Comput*, March 6, 2020. [Epub ahead of print].

[64] Cerulo, L., Tagliaferri, D., Marotta, P., Zoppoli, P., Russo, F., Mazio, C., DeFelice, M., Ceccarelli, M., and Falco, G. (2014). Identification of a novel gene signature of ES cells self-renewal fluctuation through system-wide analysis. *PLoS One*, 9 (1): e83235.

[65] Troisi, J., Landolfi, A., Sarno, L., Richards, S., Symes, S., Adair, D., Ciccone, C., Scala, G., Martinelli, P., and Guida, M. (2018). A metabolomics-based approach for non-invasive screening of fetal central nervous system anomalies. *Metabolomics*, 14 (6): 77.

[66] Attallah, O., Sharkas, M. A., and Gadelkarim, H. (2019). Fetal Brain Abnormality Classification from MRI Images of Different Gestational Age. *Brain Sci*, 9 (9): E231.

[67] Malacova, E., Tippaya, S., Bailey, H. D., Chai, K., Farrant, B. M., Gebremedhin, A. T., Leonard, H., Marinovich, M. L., Nassar, N., Phatak, A., Raynes-Greenow, C., Regan, A. K., Shand, A. W., Shepherd, C. C. J., Srinivasjois, R., Tessema, G. A., and Pereira, G. (2020). Stillbirth risk prediction using machine learning for a large cohort of births from Western Australia, 1980-2015. *Sci Rep*, 10 (1): 5354.

[68] Khatibi, T., Kheyrikoochaksarayee, N., and Sepehri, M. M. (2019). Analysis of big data for prediction of provider-initiated preterm birth and spontaneous premature deliveries and ranking the predictive features. *Arch Gynecol Obstet*, 300 (6): 1565-1582.

[69] Ramyaa, R., Hosseini, O., Krishnan, G. P., and Krishnan, S. (2019). Phenotyping Women Based on Dietary Macronutrients, Physical Activity, and Body Weight Using Machine Learning Tools. *Nutrients*, 11 (7): E1681.

[70] Huang, M. L., Hung, Y. H., and Chen, W. Y. (2010). Neural network classifier with entropy based feature selection on breast cancer diagnosis. *J Med Syst*, 34 (5): 865-873.

[71] Kadam, V.J., Jadhav, S. M., and Vijayakumar, K. (2019). Breast Cancer Diagnosis Using Feature Ensemble Learning Based on Stacked Sparse Autoencoders and Softmax Regression. *J Med Syst*, 43 (8): 263.

[72] Schaffter, T., Buist, D. S. M., Lee, C. I., Nikulin, Y., Ribli, D., Guan, Y., Lotter, W., Jie, Z., Du, H., Wang, S., Feng, J., Feng, M., Kim, H. E., Albiol, F., Albiol, A., Morrell, S., Wojna, Z., Ahsen, M. E., Asif, U., Jimeno Yepes, A., Yohanandan, S., Rabinovici-Cohen, S., Yi, D., Hoff, B., Yu, T., Chaibub Neto, E., Rubin, D. L., Lindholm, P., Margolies, L. R., McBride, R. B., Rothstein, J. H., Sieh, W., Ben-Ari, R., Harrer, S., Trister, A., Friend, S., Norman, T., Sahiner, B., Strand, F., Guinney, J., and Stolovitzky, G.; and the DM DREAM Consortium, Mackey, L., Cahoon, J., Shen, L., Sohn, J. H., Trivedi, H., Shen, Y., Buturovic, L., Pereira, J. C., Cardoso, J. S., Castro, E., Kalleberg, K. T., Pelka, O., Nedjar, I., Geras, K. J., Nensa, F., Goan, E., Koitka, S., Caballero, L., Cox, D. D., Krishnaswamy, P., Pandey, G., Friedrich, C. M., Perrin, D., Fookes, C., Shi, B., Cardoso Negrie, G., Kawczynski, M., Cho, K., Khoo, C. S., Lo, J. Y., Sorensen, A. G., and Jung, H. (2020). Evaluation of Combined Artificial Intelligence and Radiologist Assessment to Interpret Screening Mammograms. *JAMA Netw Open*, 3 (3): e200265.

[73] Bria, A., Karssemeijer, N., and Tortorella, F. (2014). Learning from unbalanced data: a cascade-based approach for detecting clustered microcalcifications. *Med Image Anal*, 18 (2): 241-252.

[74] Mordang, J. J., Gubern-Mérida, A., Bria, A., Tortorella, F., Mann, R. M., Broeders, M. J. M., den Heeten, G. J., and Karssemeijer, N. (2018). The importance of early detection of calcifications associated

with breast cancer in screening. *Breast Cancer Res Treat*, 167 (2): 451-458.

[75] Eadie, L. H., Taylor, P., and Gibson, A. P. (2012). A systematic review of computer-assisted diagnosis in diagnostic cancer imaging. *Eur J Radiol*, 81 (1): e70-e76.

[76] Arora, R., Rai, P. K., and Raman, B. (2020). Deep feature-based automatic classification of mammograms. *Med Biol Eng Comput*, March 21, 2020. [Epub ahead of print].

[77] Wei, L., Yang, Y., Nishikawa, R. M., and Jiang, Y. (2005). A study on several machine-learning methods for classification of malignant and benign clustered microcalcifications. *IEEE Trans Med Imaging*, 24 (3): 371-380.

[78] Ozcift, A. (2012). Enhanced cancer recognition system based on random forests feature elimination algorithm. *J Med Syst*, 36 (4): 2577-2585.

[79] Tanaka, H., Chiu, S. W., Watanabe, T., Kaoku, S., and Yamaguchi, T. (2019). Computer-aided diagnosis system for breast ultrasound images using deep learning. *Phys Med Biol*, 64 (23): 235013.

[80] Song, H., Li, Y., and Men, A. (2018). Microwave breast cancer detection using time-frequency representations. *Med Biol Eng Comput*, 56 (4): 571-582.

[81] Hsieh, S. L., Hsieh, S. H., Cheng, P. H., Chen, C. H., Hsu, K. P., Lee, I. S., Wang, Z., and Lai, F. (2012). Design ensemble machine learning model for breast cancer diagnosis. *J Med Syst*, 36 (5): 2841-2847.

[82] Djebbari, A., Liu, Z., Phan, S., and Famili, F. (2008). An ensemble machine learning approach to predict survival in breast cancer. *Int J Comput Biol Drug Des*, 1 (3): 275-294.

[83] Zhao, M., Tang, Y., Kim, H., and Hasegawa, K. (2018). Machine Learning With K-Means Dimensional Reduction for Predicting Survival Outcomes in Patients With Breast Cancer. *Cancer Inform*, 17: 1176935118810215.

[84] Khan, M. M., Mendes, A., and Chalup, S. K. (2018). Evolutionary Wavelet Neural Network ensembles for breast cancer and Parkinson's disease prediction. *PLoS One*, 13 (2): e0192192.

[85] Hothorn, T., Bühlmann, P., Dudoit, S., Molinaro, A., and van der Laan, M. J. (2006). Survival ensembles. *Biostatistics*, 7 (3): 355-373.

[86] Gong, I. Y., Fox, N. S., Huang, V., and Boutros, P. C. (2018). Prediction of early breast cancer patient survival using ensembles of hypoxia signatures. *PLoS One*, 13 (9): e0204123.

[87] Murata, T., Yanagisawa, T., Kurihara, T., Kaneko, M., Ota, S., Enomoto, A., Tomita, M., Sugimoto, M., Sunamura, M., Hayashida, T., Kitagawa, Y., and Jinno, H. (2019). Salivary metabolomics with alternative decision tree-based machine learning methods for breast cancer discrimination. *Breast Cancer Res Treat*, 177 (3): 591-601.

[88] Lopes, M. B., Veríssimo, A., Carrasquinha, E., Casimiro, S., Beerenwinkel, N., and Vinga, S. (2018). Ensemble outlier detection and gene selection in triple-negative breast cancer data. *BMC Bioinformatics*, 19 (1): 168.

[89] Hou, T. Y., Weng, C. F., and Leong, M. K. (2018). Insight Analysis of Promiscuous Estrogen Receptor α-Ligand Binding by a Novel Machine Learning Scheme. *Chem Res Toxicol*, 31 (8): 799-813.

[90] Martínez-Más, J., Bueno-Crespo, A., Khazendar, S., Remezal-Solano, M., Martínez-Cendán, J. P., Jassim, S., Du, H., Al Assam, H., Bourne, T., and Timmerman, D. (2019). Evaluation of machine learning methods with Fourier Transform features for classifying ovarian tumors based on ultrasound images. *PLoS One*, 14 (7): e0219388.

[91] Miao, R., Badger, T. C., Groesch, K., Diaz-Sylvester, P. L., Wilson, T., Ghareeb, A., Martin, J. A., Cregger, M., Welge, M., Bushell, C., Auvil, L., Zhu, R., Brard, L., and Braundmeier-Fleming, A. (2020). Assessment of peritoneal microbial features and tumor marker levels as potential diagnostic tools for ovarian cancer. *PLoS One*, 15 (1): e0227707.

[92] Troisi, J., Sarno, L., Landolfi, A., Scala, G., Martinelli, P., Venturella, R., Di Cello, A., Zullo, F., and Guida M. (2018). Metabolomic Signature of Endometrial Cancer. *J Proteome Res*, 17 (2): 804-812.

[93] Jiang, X., Li, J., Kan, Y., Yu, T., Chang, S., Sha, X., Zheng, H., Luo, Y., and Wang, S. (2020). MRI Based Radiomics Approach with Deep

Learning for Prediction of Vessel Invasion in Early-Stage Cervical Cancer. *IEEE/ACM Trans Comput Biol Bioinform*, January 3, 2020. [Epub ahead of print].

[94] *Baraniuk, C. (2016). Google's DeepMind to peek at NHS eye scans for disease analysis. BBC, July 6, 2016. BBC, March 14, 2020.* https://www.bbc.com/news/technology-36713308.

[95] Thompson, C. (2011). "Smarter Than You Think: What Is I.B.M.'s Watson?." *The New York Times Magazine*, February 18.

[96] Lo, Y. C., Rensi, S. E., Torng, W., and Altman, R. B. (2018). Machine learning in chemoinformatics and drug discovery. *Drug Discov Today*, 23 (8): 1538-1546.

[97] Narla, A., Kuprel, B., Sarin, K., Novoa, R., and Ko, J. (2018). Automated Classification of Skin Lesions: From Pixels to Practice. *J Invest Dermatol,* 138 (10): 2108-2110.

[98] Loeb, S. (2015). *Sense.ly and MindMeld team for AI-driven virtual nurse. June 24, 2015.* Vator, April 7, 2019. https://vator.tv/news/2015-06-24-sensely-and-mindmeld-team-for-ai-driven-virtual-nurse.

[99] Shafner, L., Abt, M., Kinch, R., Tamburri, P., Umbricht, D., and Hanina, A. (2017). Using an artificial intelligence platform on mobile devices to monitor and increase adherence in subjects with schizophrenia. Serving the needs of the diverse population. *Poster Abstract Book. American Society of Clinical Psychopharmacology Annual Meeting,* Miami, May 29-June 2, 2017: 59-60.

[100] Li, Y., Bai, C., and Reddy, C. K. (2016). A Distributed Ensemble Approach for Mining Healthcare Data under Privacy Constraints. *Inf Sci (N Y)*, 330: 245-259.

[101] Dharmarajan, K., Hsieh, A. F., Kulkarni, V. T., Lin, Z., Ross, J. S., Horwitz, L. I., Kim, N., Suter, L. G., Lin, H., Normand, S. L., and Krumholz, H. M. (2015). Trajectories of risk after hospitalization for heart failure, acute myocardial infarction, or pneumonia: retrospective cohort study. *BMJ*, 350: h411.

[102] *Obermeyer, Z., and Emanuel, E. J. (2016). Predicting the future – Big data, machine learning, and clinical medicine. N Engl J Med, 375: 1216-1219.*

[103] *Rasoini, R., Cabitza, F., Alderighi, C., and Gensini, G. F. (2017). Artificial intelligence in medicine: Between hype, uncertainty and black boxes. Toscana Medica, 11: 18-20.*

[104] *Cabitza, F., Alderighi, C., Rasoini, R., and Gensini, G. F. (2017). Unexpected potential consequences of the use of artificial intelligence systems in medicine. Recenti Prog Med, 108: 397-401.*

[105] *Chen, J. H., and Asch, S. M. (2017). Machine learning and prediction in medicine – Beyond the peak of inflated expectations. N Engl J Med, 376: 2507-2509.*

[106] *Vineis, P. (2017). Big data. Big bias. March 8, 2017. Il Pensiero Scientifico Editore, 2019.* https://pensiero.it/in-primo-piano/interviste/big-data-big-biases.

[107] *Davenport, T. H., and Glover, W. J. (2018). Artificial intelligence and the augmentation of health care decision-making. N Engl J Med Catalyst.* Massachusetts Medical Society, April 14, 2020. https://catalyst.nejm.org/doi/abs/10.1056/CAT.18.0151.

[108] *Verghese, A., Shah, N. H., and Harrington, R. A. (2017). What this computer needs is a physician. Humanism and artificial intelligence. JAMA, 319: 19-20.*

[109] *Hegel, G. W. F. (1820). Outlines of the philosophy of law. Cicero, V., Ed. Milan: Bompiani, 2006.*

[110] *Collecchia, G. (2018). Human and artificial intelligence: Cultures compared. Assist Inferm Ric, 37: 212-217.*

[111] Collecchia, G. (2013). *Artificial intelligence in medicine: Limits of big data and algorithms*. QS Edizioni, 2013. April 20, 2019. https://www.quotidianosanita.it/lettere-al-direttore/articolo.php?articolo_id=73313.

[112] Cabitza, F., Rasoini, R., and Gensini, G. F. (2017). Unintended Consequences of Machine Learning in Medicine. *JAMA*, 318 (6): 517-518.

[113] Povyakalo, A. A., Alberdi, E., Strigini, L., and Ayton, P. (2013). How to discriminate between computer-aided and computer-hindered decisions: a case study in mammography. *Med Decis Making*, 33 (1): 98-107.

[114] Turley, A., Roberts, A., Evemy, K., Haq, I., Irvine, T., and Adams, P. (2007). Diagnostic accuracy of automated computerized electrocardiogram interpretation compared with a panel of experienced cardiologists. *Critical Care*, 11 (Suppl 2): S100.

[115] Caruana, R., Lou, Y., Gehrke, J., Koch, P., Sturm, M., and Elhadad, N. (2015). Intelligible Models for Healthcare: Predicting Pneumonia Risk and Hospital 30-day Readmission. *KDD*, August 10-13, 2015, Sydney, NSW, Australia.

[116] Stanescu, A., and Pandey, G. (2017). Learning parsimonious ensembles for unbalanced computational genomics problems. *Pac Symp Biocomput*, 22: 288-299.

[117] Roccetti, M., Prandi, C., Salomoni, P., and Marfia, G. (2016). Unleashing the true potential of social networks: confirming infliximab medical trials through Facebook posts. *Netw Model Anal Health Inform Bioinforma*, 5: 1-13.

[118] Saeb, S., Lonini, L., Jayaraman, A., Mohr, D. C., and Kording, K. P. (2017). The need to approximate the use-case in clinical machine learning. *Gigascience*, 6 (5): 1-9.

[119] Little, M. A., Varoquaux, G., Saeb, S., Lonini, L., Jayaraman, A., Mohr, D. C., and Kording, K. P. (2017). Using and understanding cross-validation strategies. Perspectives on Saeb et al. *Gigascience*, 6 (5): 1-6.

[120] DeMasi, O., Kording, K., and Recht, B. (2017). Meaningless comparisons lead to false optimism in medical machine learning. *PLoS One*, 12 (9): e0184604.

[121] Rui, W., Chen, F., Chen, Z., Li, T., Harari, G., Tignor, S., Zhou, X., Ben-Zeev, D., and Campbell, A. T. (2014). StudentLife Assessing Mental Health, Academic Performance and Behavioral Trends of College Students using Smartphones. In: *Proceedings of the ACM Conference on Ubiquitous Computing*.

[122] Aharony, N., Pan, W., Ip, C., Khayal, I., and Pentland, A. (2011). SocialfMRI: Investigating and shaping social mechanisms in the real world. *Pervasive and Mobile Computing*, 7 (6): 643-659.

[123] Molinari, R., and Roccetti, M. (2018). *Some cases of misleading results in the application of machine learning in medicine.* Alma Mater Studiorum, Università di Bologna, Scuola di Scienze, Corso di Laurea in Informatica per il Management, Sessione II, 2017/2018. https://amslaurea.unibo.it/16758/1/molinari_riccardo_tesi.pdf.

[124] Yu, X. T., and Zeng, T. (2018). Integrative Analysis of Omics Big Data. *Methods Mol Biol*, 1754: 109-135.

[125] Cascianelli, S., Scialpi, M., Amici, S., Forini, N., Minestrini, M., Fravolini, M. L., Sinzinger, H., Schillaci, O., and Palumbo, B. (2017). Role of Artificial Intelligence Techniques (Automatic Classifiers) in Molecular Imaging Modalities in Neurodegenerative Diseases. *Curr Alzheimer Res*, 14 (2): 198-207.

[126] Van Poucke, S., Thomeer, M., Heath, J., and Vukicevic, M. (2016). Are Randomized Controlled Trials the (G)old Standard? From Clinical Intelligence to Prescriptive Analytics. *J Med Internet Res*, 18 (7): e185.

[127] Mazzaracca, R. (2019). *Machine learning promises to revolutionize medicine*. Rarelab, June 10, 2019. https://www.osservatorio terapieavanzate.it/news/il-machine-learning-promette-di-rivoluzionare-la-medicina.

[128] (2018). *Machine learning applications in medicine*. WordPress, March 30, 2020. http://www.starparty.it/le-applicazioni-del-machine-learning-nella-medicina/.

[129] (2020). *Flagler Hospital uses AI to create clinical pathways that enhance care and slash costs*. Ayasdi A LCC, March 2, 2020. https://www.ayasdi.com/company/news-and-events/news/10194/.

In: An Introduction to Approaches … ISBN: 978-1-53618-680-2
Editor: Yi-Tung Chan

Chapter 8

ENSEMBLE LEARNING APPROACH IN AUTOMATED MODAL IDENTIFICATION

Mahdi Ghamami[1,*], Hassan Nahvi[1] and Vahid Yaghoubi[1,2]
[1]Department of Mechanical Engineering, Isfahan University of Technology, Isfahan, Iran
[2]Department of Materials, Textiles and Chemical Engineering, Ghent University, Zwijnaarde, Belgium

ABSTRACT

A modal identification is a type of system identification that uses modal analysis to extract the dynamic parameters of unknown systems. Machine learning enables the automation of the identification process so that the intelligent algorithm is replaced by the human decision maker. To enhance the robustness of the identification algorithm and extend the ability to tackle more complex issues from the viewpoint of system identification, the algorithm was developed using ensemble learning approach and a modified algorithm was presented. Several machine learning algorithms such as FCM, DBSCAN, AGNES, LIDA, KNN and,

* Corresponding Author's Email: mghamazi@ut.ac.ir.

etc. were used as a decision maker and the fusion among decision makers was made possible through one of the data fusion methods called Dempster–Shafer theory. This theory operates by the probabilistic basis and provides a base for the operation of different machine learning algorithms and the combination between their decisions. The results are very promising and the modified algorithm provides more accurate and robust results that is particularly important in industrial issues in the areas of system identification, condition monitoring, fault detection and damage localization which are commonly seen in mechanical, aerospace and civil engineering. The proposed identification algorithm deals with many different fields in statistics, mechanical and computer engineering such as data acquisition, signal processing, resampling, model reduction, datamining, parameter estimation, feature extraction, classification, clustering, data fusion and automation. This chapter focuses on the ensemble learning aspect exploring its future prospects. Two case studies have been investigated and both analytical and experimental aspects have been considered. The results show the modified algorithm is of great ability to separate modal parameters precisely.

Keywords: ensemble learning, modal identification, data fusion, bootstrap, Dempster–Shafer theory

1. INTRODUCTION

Two different areas including machine learning and modal identification are involved in this field. Machine learning is a subcategory of Artificial Intelligence that focuses mainly on developing logical computer programs that can access data (experience) and use it to improve themselves without being explicitly programmed. Ensemble learning is a method that combines diverse learning algorithms together to improve the stability and predictive power of the meta-model more than it could be obtained from any of the constituent learning algorithms alone. It uses several machine learning models built of different learning algorithms to improve the prediction accuracy. Ensemble learning is always used in our daily lives, for example, if it is possible, we prefer to know the advice of several doctors from different fields when we have the symptoms of the disease, also buyers prefer to know the price of a desired product by several sellers instead of buying the product from a specific store. That's why we have several representatives who decide in parliament. Different representatives are like different learning models that process data differently. Combining the decision of different models, as long as they are relatively independent and diverse (like good politicians), we will reach a better final result than the result of each individual decision.

Modal identification uses modal analysis to derive the dynamic properties (modal parameters) of systems. Modal parameters such as natural frequencies, damping ratio and mode shapes are important because they can describe the dynamic behavior of system. The algorithm is a "fully automatic" type which its identification ability is generalizable to other systems when it has been done automatically and there is no need to the prior information or the human decision maker. In recent years, smart structures have attracted much interest which have many applications in practical issues. On the other hand, inverse problem theory provides methods for the practical solution of applied problems. These methods usually need only one dataset for identification process, called initial dataset. In modal identification, the initial dataset measures vibrational signals which are collected in a modal testing without repetition of data acquisition. Under noisy condition, the modal parameters cannot be accurately estimated. For this reason, machine learning methods can be very helpful and effective.

2. Background

Ensemble learning in modal identification has two important aspects: the first is the modal identification, which is a subset of system identification, and the second is the ensemble learning, which is a subset of artificial intelligence. Few studies have been conducted in this field so that both aspects have been considered. In order to investigate the challenges of this field, some related studies have been reviewed in the following. Modal-based algorithms are well-known techniques for identification, damage detection, tracking and monitoring in frequency, modal or time domain [1-9]. But the lack of automated algorithms has long been a limitation in their widespread use. In recent years, many algorithms have been introduced and some of them use statistical techniques for identification. These algorithms use various methods to eliminate the need of threshold and parameter definition by user. In order to overcome this limitation, a number of "fully automated" algorithms have been proposed in recent years. Artificial intelligence and machine learning algorithms play an effective role in automation of identification algorithms. Most automated modal identification algorithms can only carry out a part of the identification process automatically, while a few of them are capable to fully automated modal identification [10-13]. General structure of semi-automated and fully automated modal identification algorithm is illustrated in Figure 1.

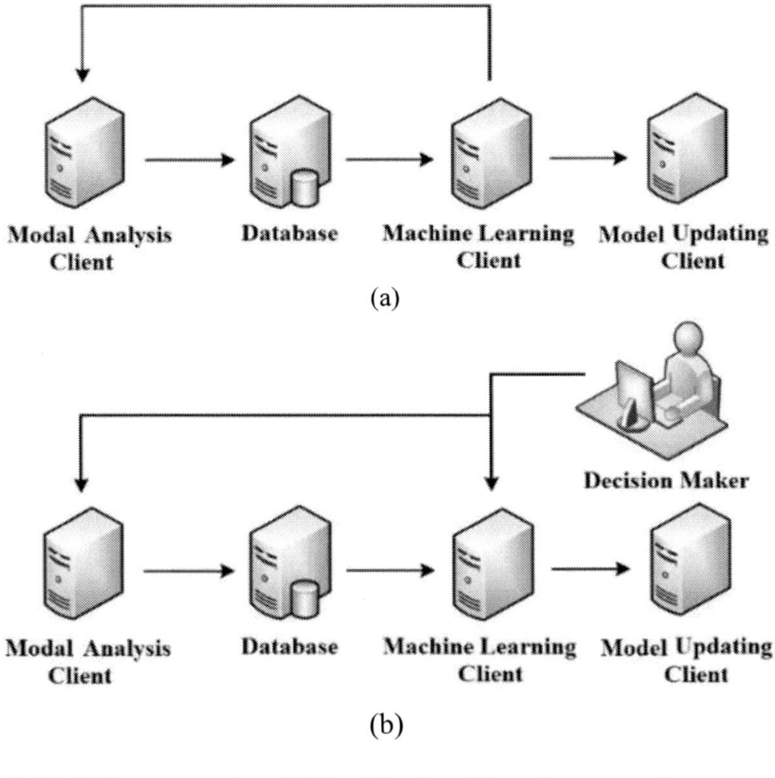

Figure 1. (a) Fully automated modal identification, (b) semi-automated modal identification.

Some researchers have investigated the challenges in this field such as close modes [14-16], high number of data-set [17], damage localization [18], long-term monitoring [19-21], tensile load estimation [22], accurate damping estimation [23], low signal to noise ratio [14], and damped modes [4]. Several other innovative applications were developed in regard to modal identification of large systems [20-21], Scanning Laser Vibrometry and uncertainties [24-25] and model updating [26-27]. In condition monitoring, there is a possibility to compare between the various datasets and there is prior information of the system [28-30], while modal identification algorithms usually have access only to a dataset and there is no prior information [31-32]. The need for numerous initial dataset is the common assumption in most of aforementioned studies. Recently studies show that interest for fully automated methods that are generalizable to other systems have increased. Automated identification algorithms require adequate variety of data to perform the identification process correctly. This seeming contradiction leads to ensemble learning.

The theoretical result in ensemble learning leads to that many decorrelated “weak learners” can asymptotically form a “strong learner”. The combination will do better than any algorithm alone although each of them might perform well individually. Typically, this happens because each individual model might overfit to a different part of the data. By combining different individual models into an ensemble, their individual mistakes can be averaged out to reduce the risk of over-fitting while maintaining strong prediction performance. Many researches have been done on ensemble learning field which include: Multiple classifier [33-35], Combining classifier [36], Dynamic classifier selection [37-38], Image classification [39], Malware detection [40-41], Face recognition [42-44], Emotion recognition [45-49], Fraud detection and Financial [50-52], Computer security [53], Medicine [54-58]. Based on the above, it can be seen that both fields are very innovative and many studies are underway. In the following, the theory and required explanations of the desired algorithms are presented.

3. DISSEMINATION AND THEORETICAL CONSIDERATIONS

In this section a very quick overview and the basic equations of related theories will be summarized. The development of these equations can be found in detail in the references.

3.1. Ensemble Learning

Ensemble methods are meta-algorithms that combine several machine learning techniques into one predictive model in order to decrease variance (bagging), bias (boosting), or improve predictions (stacking). To understand what really goes behind an ensemble model, we need to first understand what causes error in the model. The error emerging from any model can be broken down into three components mathematically (Eq 1). Following are these component:

$$Error(x) = \left(E[\hat{f}(x)] - f(x)\right)^2 + E\left[\hat{f}(x) - E[\hat{f}(x)]\right]^2 + \sigma_e^2 \quad (1)$$
$$Error(x) = Bias^2 + Variance + Irreducible\ Error$$

Error can be divided into the following two types: Reducible and Irreducible error. Reducible errors including bias and variance can be reduced by improving the model while Irreducible errors cannot be reduced at all. Variance is also known as random error, random variation, or noise. On the other hand, bias or Systematic error refers to deviations that are not due to chance alone. The simplest example occurs with a measuring device that is improperly calibrated so that it consistently overestimates (or underestimates) the measurements.

Variance has no preferred direction, so we expect that averaging over a large number of observations will yield a net effect of zero. The estimate may be imprecise, but not inaccurate. The impact of random error, imprecision, can be minimized with large sample sizes. Bias, on the other hand, has a net direction and magnitude so that averaging over a large number of observations does not eliminate its effect. In fact, bias can be large enough to invalidate any conclusions. Increasing the sample size is not going to help. Bias can be subtle and difficult to detect. Variance corresponds to imprecision, and bias to inaccuracy.

A complete model should maintain a balanced reduction between these two types of errors. This is known as the trade-off management of bias-variance errors. Ensemble learning is one way to execute this trade-off analysis. Most ensemble methods use a single base learning algorithm to produce homogeneous base learners. The learners of the same type leading to "homogeneous ensembles". There are also some methods that use heterogeneous learners which leading to "heterogeneous ensembles". In order for ensemble methods to be more accurate than any of its individual members, the base learners have to be as accurate as possible and as diverse as possible. Ensemble Learning can be divided in three groups:

- Sequential Ensemble Methods: these methods popularly known as "boosting" which the weak learners are generated sequentially. The basic motivation of sequential methods is to exploit the dependence between the base learners. The overall performance can be boosted by weighing previously mislabeled examples with higher weight. Boosting can be used primarily for reducing bias and also variance. It is a weighted average approach by iteratively learning and forming multiple weak learner and combining them to make a final strong learner. Boosting is an interesting approach but is very noise sensitive and is only effective using weak classifiers.
- Parallel Ensemble Methods: these methods popularly known as "bagging" which the weak learners are generated parallel. The basic motivation of parallel methods is to exploit independence between the base learners since the error can be reduced dramatically by averaging. Bagging helps when it is applied to an over fitted model (low bias high variance) and it can be used for reducing variance while keeping the bias simultaneously. It does not help much with models which have high bias. One way to reduce the variance of an

estimate is averaging multiple estimates. Bagging uses averaging for regression.

- Stacked Ensemble Methods: these methods popularly known as "stacking" that often considers heterogeneous evidences (base learners, weak learners), learns them in parallel and combines them by training a meta-model to output a prediction. This is a very interesting way of combining models. This can lead to decrease in either bias or variance error depending on the combining learner. It has also been used to estimate bagging's error rate [59-60]. It has been reported to outperform Bayesian model-averaging [61]. Two things are needed to build the stacking model: the learners and the meta-model (final predictor) that combines them.

General structure of ensemble learning types are illustrated in Figure 2.

If the algorithm faces imbalanced datasets, noisy datasets, ill-defined problems, the use of ensemble learning can be more acceptable. In imbalanced datasets, few samples are in one class (minority class), and many samples are in other classes (majority classes). Noisy datasets have the outlier or unquantified uncertainty and in ill-defined problems, there are few data with poor quality. If all evidences misclassify the same samples, the performance won't increase. In order to increase performance, the evidences in an ensemble need to correctly classify different samples and they need to be diverse.

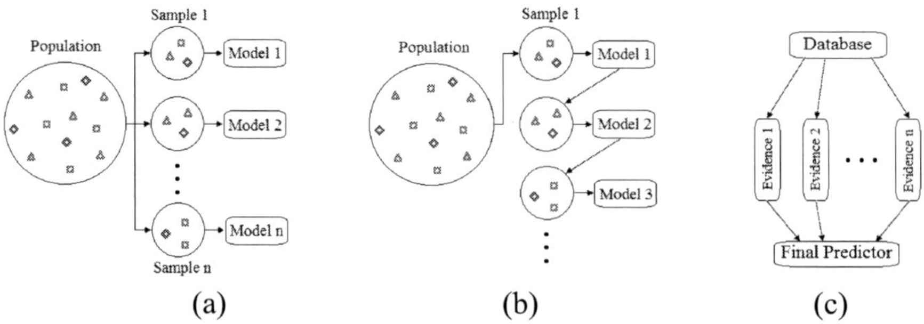

Figure 2. (a) Bagging, (b) Boosting, (c) Stacking.

3.2. Dempster-Shafer Theory

Meta-model in ensemble learning can use of Bayesian approach, data fusion theory and, etc. Data fusion theory, like artificial neural network, fuzzy logic and other methods of artificial intelligence, is inspired by the function of the human brain. Data fusion deals with combining data from different sensors and evidence to predict system properties. Data fusion takes place at three levels: data (signal) level, feature level, and evidence (decision maker) level. Using a multi-sensor data fusion strategy can increase system performance. Increasing the degree of reliability, reducing the noise level, reducing the outlier data are some of the advantages of data fusion theory. Dempster-Shafer's theory or in short DST provided an analysis of properties being desirable for subjective measures of uncertainty. The theory is given by Arthure P. Dempster in 1967 and his student Glenn Shafer in 1976. This theory is being released because of two reason: 1) Bayesian theory is only concerned about single evidences. 2) Bayesian probability cannot describe ignorance. DST is an evidence theory, it combines all possible outcomes of the problem. DST is one of the most powerful methods of data fusion and deals with fusion at the decision level. Increasing decision maker, while increasing computation, will not result in confusion. Shafer completed this theory and corrected its shortcomings, expanding the theory to analyze imprecise and incomplete information. Hence it is used to solve problems where there may be a chance that a different evidence will lead to some different result. The uncertainty in this model is given by:

- Consider all possible outcomes.
- Belief will lead to believe in some possibility by bringing out some evidence.
- Plausibility will make evidence compatibility with possible outcomes

In traditional probability theory we have only one probabilities, but in evidence theory we are dealing with a set of probabilities. Evaluating evidences from subjects may face to highly variable states, for example, two evidences may be perfectly inclusion, inclusion-integration, integration, or conflict. The four general states which can be happened between evidences in DST are shown in Figure 3. In this figure, the evidence is displayed by A to E. The occupied area by each evidence indicates evidence decision that can be in showed states. Accordingly, these four different states may occur among the evidence. A and D states are rare and B and C states are more common in practical issues.

The advantages of DST include the following: 1) As we add more information, uncertainty interval reduces. 2) DST has much lower level of ignorance. 3) Diagnose Hierarchies can be represented using this. Also, the disadvantages of the DST include its computational cost. In this computation effort is high, as we have to deal with 2^n of sets. The classical DST is a theory which is based on frame of discernment. The frame of discernment is composed of n exhaustive and exclusive hypotheses [62]. The Basic Probability Assignment (BPA) for all hypotheses must satisfy the following conditions:

$$0 \leq m(A_i) \leq 1, \qquad m(\Phi) = 0, \qquad \sum m(A_i) = 1 \tag{2}$$

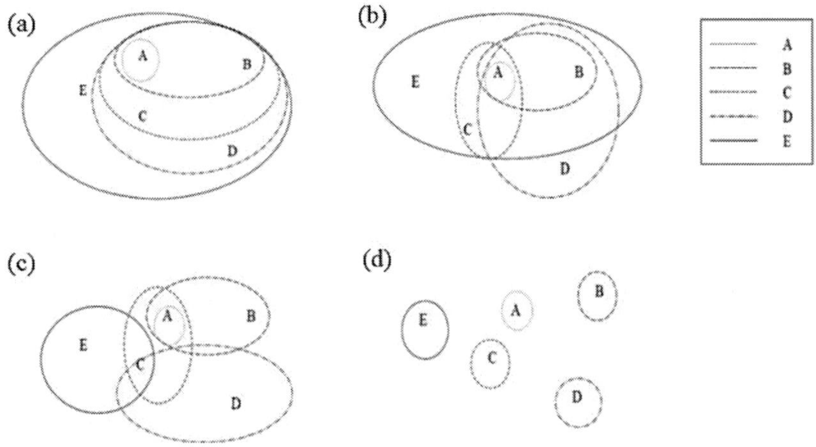

Figure 3. The four general states in DST, (a) inclusion, (b) inclusion-integration, (c) integration, (d) conflict.

Definitions such as belief function and plausibility function and so on have been discussed in related studies [62]. It do not need to present in this study again. DST is an efficient theory which offers appropriate aggregation method for BPAs. The combination rule of DST assumes the sources of evidences are independent. The classical combination rule of DST is defined as:

$$m(A) = \frac{\sum_{A_i \cap B_j = A} m_1(A_i) m(B_j)}{1 - \sum_{A_i \cap B_j = \Phi} m_1(A_i) m_2(B_j)}, \qquad A \neq \Phi \tag{3}$$

Where, A and B are focal target elements. In Eq (3), the term of $\sum_{A_i \cap B_j = \Phi} m_1(A_i) m_2(B_j)$ known as "K". K denotes a measure of conflict between sources of evidences. The denominator 1-K is addressed normalization factor. The larger K is, the higher the conflict of sources between evidences exists [63]. If K = 1, the combination rule cannot be used to fuse and the orthogonal sum does not exist as well. The combination rule has two properties: commutativity and associativity. One difference in perspective is that Bayesian analysis focuses on assigning nonnegative weights (typically probabilities) to each individual proposition forming a set of mutually exclusive propositions. Alternatively, DST assigns nonnegative weights (mass) to each "combination" of events. This allows for the inclusion of information with "unquantified uncertainty", and DST can be considered a generalization of Bayesians. DST is more akin to rough sets than fuzzy logic, also it is different from game theory. Game Theory is the study of mathematical models of conflict and cooperation between intelligent rational decision-makers. DST is the mathematical study of strategies for optimal decision-making between options involving different risks or expectations of gain or loss depending on the outcome.

3.3. Modal Identification

Modal identification is the core of proposed system identification. The rest of the system identification process are matrix operations, sorting and optimization which relate smaller algorithms to each other. The ultimate goal of the proposed algorithm is separating of physical modes from spurious modes. Physical modes are vibrational modes belonging to the system, while spurious modes are unrealistic modes and not related to the system and appear in results due to computational error. Various steps of modal identification are shown in Figure 4.

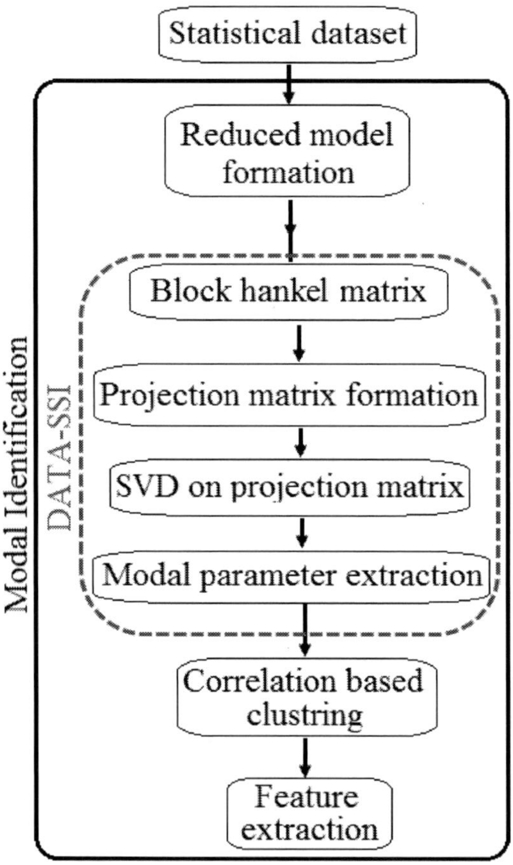

Figure 4. Automated modal identification algorithm.

The modal identification process begins with exhaustive model which is a state-space model with high model order. High model order will increase computational cost. With the help of the model-reduction method, the model order of the exhaustive model has been reduced to the extent possible. This causes spurious modes are removed as much as possible and at the same time, the physical modes are not eliminated from the exhaustive model. The model-reduction method is an optimization algorithm. The optimization algorithm minimize the SVC which is a criteria based on the exhaustive model (Eq 4).

$$SVC(n) = \sigma_{n+1}^2 + \frac{C(N_d)d(n)}{N_d} \qquad n = 1,2, \ldots 2n_e \tag{4}$$

In the exhaustive model, the model order is considered equal to $2n_e$. σ_{n+1} is the estimated Hankel singular values of the system's balanced Gramian, *d(n)* is the minimum number of parameters of a state-space model by order *n*, N_d is the number of given data points and $C(N_d)$ is a model complexity penalty factor. After this optimization, a model order is achieved which state-space model based on this model order is known as the reduced model and the continuation of the identification process is done based on the reduced model.

The Data-driven Stochastic Subspace Identification algorithm (DATA-SSI) is the most important step of the modal identification algorithm. DATA-SSI extracts the modal parameters of all statistical resamples with the help of the estimated model order. The equation of motion for a simple discrete mechanical system with viscous damping can be written as follows:

$$[\mathrm{M}]\{\ddot{x}(t)\} + [\mathrm{C}]\{\dot{x}(t)\} + [\mathrm{K}]\{x(t)\} = \{f(t)\} \tag{5}$$

[M] is Mass matrix, [C] is viscous damping matrix, [K] is Stiffness matrix, $\{x(t)\}$ is displacement vector, $\{f(t)\}$ is external force vector and t is time. The DATA-SSI assumes the linear and time-invariant (LTI) system which can be represented in a state-space form as Eq (6):

$$\mathrm{X(t+1) = [A]X(t) + [B]F(t) + w(t)\,; \quad Y(t) = [C]X(t) + [D]F(t) + v(t)} \tag{6}$$

The Block Hankel matrix $[\mathrm{Y}_h]$ defined in SSI is simply a gathering of a family of matrices that are created by shifting the data matrix:

$$[\mathrm{Y}_h] = \begin{bmatrix} [\mathrm{Y}_{(1:\mathrm{N}-2\mathrm{L})}] \\ [\mathrm{Y}_{(2:\mathrm{N}-2\mathrm{L}+1)}] \\ \vdots \\ [\mathrm{Y}_{(2\mathrm{L}:\mathrm{N})}] \end{bmatrix} = \begin{bmatrix} [\mathrm{Y}_{\mathrm{hp}}] \\ [\mathrm{Y}_{\mathrm{hf}}] \end{bmatrix} \tag{7}$$

Whereas Y displays system response. The projection matrix is constructed based on the projection of the future unto the past defines the matrix like Eq (8):

$$\mathrm{O} = \mathrm{E}([\mathrm{Y}_{\mathrm{hf}}][\mathrm{Y}_{\mathrm{hp}}]) \tag{8}$$

By using SVD on the projection matrix, initial conditions and observability matrix can be estimate by Eq (9):

$$[\mathrm{O}] = [\mathrm{U}][\mathrm{S}][\mathrm{V}]^T, \ [\mathrm{O}] = [\hat{\Gamma}]X_0 \Rightarrow [X_0] = [\mathrm{S}]^{1/2}[\mathrm{V}]^T, \ [\hat{\Gamma}] = [\mathrm{U}][\mathrm{S}]^{1/2} \tag{9}$$

The system matrix $[\hat{A}]$ and the observation matrix $[\hat{C}]$ can be found from the estimate of the matrix $[\hat{\Gamma}]$:

$$[\hat{\Gamma}_{(2:\mathrm{L})}][\hat{A}] = [\hat{\Gamma}_{(1:\mathrm{L}-1)}], \quad [\hat{C}] = [\hat{\Gamma}_{(1:1)}] \tag{10}$$

Modal parameters can be found by an eigenvalue decomposition of the system matrix $[\hat{A}]$:

$$[\hat{A}] = [\Psi][\mathrm{V}_k][\Psi]^{-1} \tag{11}$$

Modal parameters can be extracted by using the below formules (Eq 12):

$$(\lambda_k) = \frac{Ln\,(V_k)}{\Delta\mathrm{T}}, \quad \omega_k = |\lambda_k|, \quad (\xi_k) = \frac{Re\,(\lambda_k)}{|\lambda_k|}, \quad [\Phi] = [C][\Psi] \tag{12}$$

ω_k is natural frequency, ξ_k is damping ratio and $[\Phi]$ is the mode shape matrix. By expanding the modal parameters, the last step of the SSI algorithm ends. In order to find further information and mathematical relationships of this algorithm, references [64-66] are very useful. After extraction of modal parameters (eigenfrequency, damping ratio, mode shape matrix) from the statistical dataset, the classification step starts by a correlation analysis algorithm which is a non-iterative process and categorizes similar modes. Two metrics, higher order Modal Observability Correlation (*hMOC*) and basic Modal Observability Correlation (*bMOC*) are used to compare the similarity of the extracted modes with each other [67]. These metrics are shown in Eq (13):

$$hMOC(u,\bar{o}_k) = \cos^2(\measuredangle[u,\bar{o}_k]),\ bMOC(\bar{o}_i,\bar{o}_j) = \frac{\left|\bar{o}_i^{\mathrm{H}}\bar{o}_j\right|^2}{(\bar{o}_i^{\mathrm{H}}\bar{o}_i)(\bar{o}_j^{\mathrm{H}}\bar{o}_j)} \quad i,j = 1,2,\ldots,n_x \tag{13}$$

In above Equation, $\measuredangle$ denotes the angle between the vector $\bar{o}_k$ and the subspace u. The vectors $\bar{o}_i$ and $\bar{o}_j$ contain information about the eigenvectors and the associated eigenvalues [67]. Three features are defined for each class. Each method will be briefly described and its position will be explained in the process. Eager readers can refer to the cited references for more information. Feature, is a criterion for each cluster in the likely physical class that make distinction between clusters. These features are used by the clustering algorithm in order to clusters separation. The features are shown in Eq. (14):

$$\chi_1^i = N_{m_i} log(MC_i^b) \quad \chi_2^i = log((\frac{\sqrt{Var\ (\omega_i)}}{\mu(\omega_i)})^{-1}) \ \chi_3^i = \mu^{-1}(\xi_i) \tag{14}$$

The above features must be normalized such that the aligned features must have the same range. All features are scaled to the interval [0, 1] if they tend to unity, they represent a high Degree of Physicalness (DoP). Else if they tend to zero they represent low DoP and related to the spurious modes [12]. The feature space can be used by clustering algorithms, clustering algorithms try to identify the type of cluster and categorize it into one of two physical or spurious class by using the values of the assigned feature to each cluster. Because the initial signals are generated by statistical resampling, the variance of cluster can be considered as uncertainty quantification.

3.4. Uncertainty

The result of any physical measurement has two essential components: 1) A "numerical value" (in a specified system of units) giving the best estimate possible of the quantity measured. 2) the "degree of uncertainty" associated with this estimated value. Uncertainty is a parameter characterizing the range of values within which the value of the measurand can be said to lie within a specified level of confidence. The uncertainty is a quantitative indication of the quality of the result. Uncertainty may be reduced, but never eliminated. There are measures for estimating uncertainty, such as standard deviation that are based entirely on the analysis of experimental data when all of the major sources of variability were sampled in the collection of the dataset.

4. Ensemble Modal Identification

Using the above mentioned theories, the ensemble modal identification algorithm is proposed as shown in Figure 5. These is a meta-algorithms which combines several machine learning techniques into one predictive model in order to decrease the variance and bias. First, the initial dataset is measured by the modal testing, second, the modal parameters are extracted by the modal identification algorithm, third, the same modal parameters are collected in separate clusters, fourth, the features of each cluster are extracted and the features database is completed. Fifth, several evidences simultaneously examine the space of features and classify mode shapes. Sixth, the fusion algorithm combines the decision of different evidence and presents the final result for each mode shape.

Ensemble approach is combining many different independent machine learning models to arrive at final decision. This decision is taken by majority vote of these models. Each of the constituent models can be just better than a random guess. Every evidence consists of two steps: 1) Producing a distribution of simple ML models on subsets of the original data. 2) Combining the distribution into one "aggregated" model. The independency here refers to the statistical independence of the probability spaces generated by classifiers. Statistical independence occurs when the probability of the results of one classification does not affect the other.

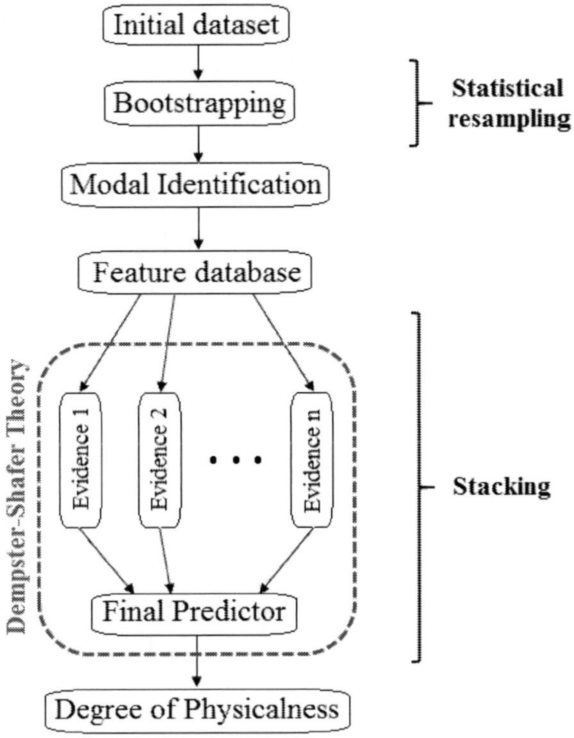

Figure 5. Ensemble automated modal identification algorithm.

DST is a method by which a large number of weak/moderate learners (or learning algorithms) can merge to produce a better and more robust decision maker. It can yield superior decision making ability. This approach works well when the base-level algorithms are unstable: they tend to produce different representations depending on which subset of the data is chosen. Decision trees, SVM (Support Vector Machine), KNN (K-Nearest Neighbors algorithm) and ANNs (Artificial Neural Network) are unstable, but linear classifiers tend to be stable and so would not work well with ensembles. For proposed ensemble algorithm, weak learners such as KNN, FCM (Fuzzy C-Means clustering), DBSCAN (Density-Based Spatial

Clustering of Applications with Noise), AGNES (Agglomerative Nesting) and LIDA (Learning Intelligent Distribution Agent) are used as evidence in DST.

5. Case Studies

In this section, applications of ensemble modal identification algorithm are investigated by two case studies. The first case study is a concrete arch dam, which is one of the large scale structures in the field of civil engineering, and the second case study is a cooling fan, one of the most widely used components in the industries.

5.1. Concrete Arch Dam

Dam safety is a major challenge for owners since the failure of dams can cause economic, environmental and social consequences to the downstream population and the environment. Dam safety monitoring program is an inevitable part of evaluating dam safety status. From the condition monitoring point of view, modal parameters have particular importance in damage assessment and detection. For this purpose, Ensemble modal identification algorithm is performed to identify modal parameters of an arch dam. For this purpose, Morrow point arch dam is selected and a Finite Element (FE) of the dam is established. In the first step, the natural frequencies of the dam based on the forced vibration tests are utilized to verify the developed finite-element model of the dam. Afterwards, the dam is excited by El-Centro earthquake and its seismic response is recorded. The collected signals are presented to the proposed algorithm. The results indicate acceptable identification by the proposed algorithm using a dataset and non-repetitive data acquisition.

The Morrow point dam (Figure 6) is an approximately symmetric arch dam located on the Gunnison River in Colorado. It is 144 m in height with length and thickness of 219 m and 3.7 m in the crest, respectively, and 16 m base thickness. A FE model of the dam is developed, as shown in Figure 6. In this case, the dam body is modeled by eight-node brick elements (C3D8R) which, in turn, the currently developed model consists of 1024 nodes. It is noteworthy to state that the linear behavior is assumed for the dam material. The concrete has a modulus of elasticity E_c =34.5 Gpa. In addition, the mass density of concrete and Poisson's ratio are considered to be ρ_c= 2403 kg/m^3, and ν_c =0.2, respectively.

Figure 6. The Morrow point dam (left) FE model and measurement points on down-stream face of the dam (right).

The frequency range of 0 to 10 Hz has been investigated and six vibrational modes are observed in this range. Figure 6 shows the well-instrumented strong motion array on the down-stream face of the dam that record seismic response data. The first six mode shapes and corresponding natural frequencies of the dam are presented in Figure 7.

The natural frequencies of the Morrow point dam based on the FE analysis and the forced vibration field tests for three different years from 1969 to 1985 with different water levels below the crest 48m, 3m, and 2m reported by John F. Hall are tabulated in Table 1. It should be noted that the highest and the lowest percentage of difference is assigned to the symmetric type of the third mode and the first mode, being approximately equal to be 16.4% and 0.2%, respectively. Also, the average amount of difference percentage is about 5.88%. The likely cause for this amount of difference is merely modeling the dam without considering the reservoir modeling.

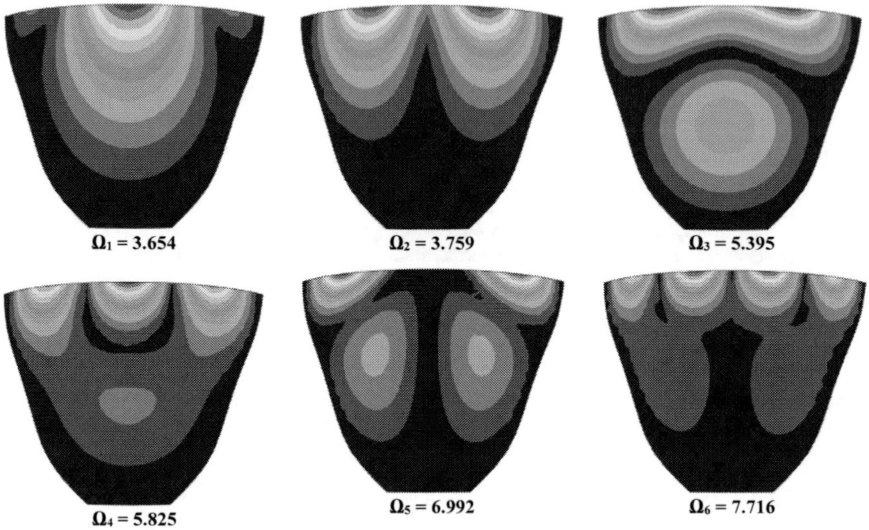

Figure 7. The first six mode shapes and corresponding natural frequencies of the dam.

After FE model verification, the modal parameters of the dam are extracted based on initial dataset (measured signals). The operational vibrational signal can be easily derived from seismic response. For the purpose of seismic analysis of the Morrow point dam, the horizontal component of 1940 El-Centro is selected and applied at the abutment of the dam. Figure 8 shows the time history ground acceleration.

Table 1. Comparisons between natural frequencies (Hz) obtained from forced vibration field tests [68] and finite element model of Morrow point dam

mode	type	Forced vibration test in			Present FEM	Difference (%)
		1969	1972	1985		
1	Symmetric	3.67	-	2.95	3.654	0.2
2	Antisymmetric	3.73	3.21	3.30	3.759	1
3	Symmetric	4.52	3.93	3.95	5.395	16.4
4	Symmetric	-	-	5.40	5.825	-
5	Antisymmetric	6.67	6.05	6.21	6.992	4.7
6	Symmetric	7.18	6.73	6.70	7.716	7.1

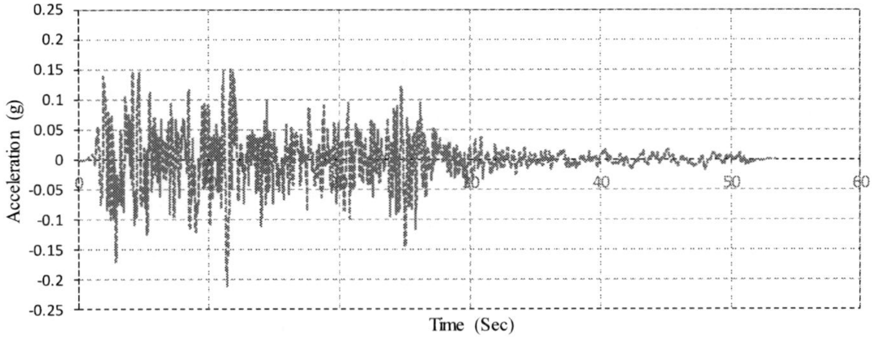

Figure 8. Time history of El-Centro acceleration.

Five collected signals which make the initial dataset are presented to the proposed identification algorithm. The sampling frequency and signal length are chosen based on the frequency range. The discrete-time signal and its frequency component, which are measurements of the amplitude at various frequencies, are presented. Velocity time-domain record and corresponding Power Spectral Density (PSD) at the first measurement node are shown in Figure 9.

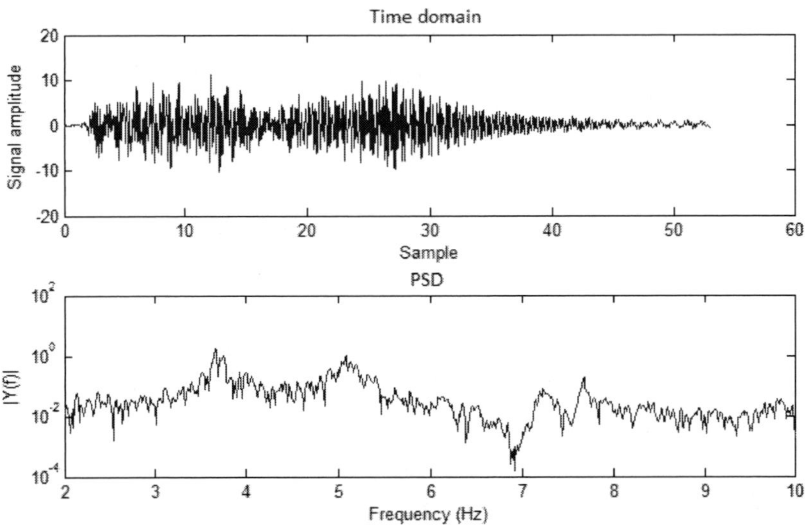

Figure 9. Measured signal of the first measurement point, Time domain (up), Frequency domain (down).

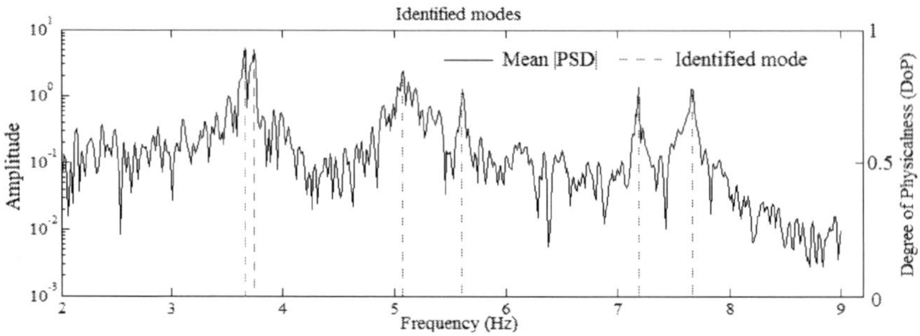

Figure 10. Identified physical modes of the dam.

Table 2. Values of identified modes of the dam

Mode	1st	2nd	3rd	4th	5th	6th
FEM (Hz)	3.65	3.75	5.39	5.82	6.99	7.71
IA (Hz)	3.66	3.74	5.07	5.61	7.19	7.66

The modes appear when the frequency component are plotted and the human user can easily identify the peaks. Nevertheless, the system identification algorithm only deals with numerical data and is incapable of identifying like users. To separate the physical modes from spurious modes, 30 statistical signals are generated from each signal with the help of Bootstrapping. Signal variety in the initial dataset increases to 150 statistical signals and the identification process is performed. The likely physical modes identified by the algorithm are presented in Figure 10. It shows the mean of the PSDs in the initial dataset, and the identified physical modes of dam. According to the shown diagram, the authenticity of the proposed algorithm in identifying system modes is crystal clear.

As mentioned, there are six vibrational modes in a range of 0 to 10 Hz, and the algorithm has succeeded in identifying all the vibrating modes in this range. They are identified by the Identification Algorithm (IA) as shown in Figure 10. Table 2 presents the natural frequency of dam based on the FE and IA methods.

The results listed in the table represent that the automated algorithm is able to identify the modal parameters very well. It should be noted that the maximum difference between the results of two methods is less than 2.12%. Also, for the aforementioned results, the average difference doesn't exceed 0.24%. The computation of the modal parameters took less than 20 seconds. An extensive validation between FEM and the proposed identification algorithm results illustrated the accuracy and robustness of the automated identification method.

5.2. Cooling Fan

In previous case study, identification algorithms can easily identify single (well-separated) modes. Another important challenge in system identification field is the close modes and multiple modes. In theory, double or multiple modes may occur but in the real world, such modes appear as close mode or high-density region of modes in frequency domain. Identifying such modes is very important and requires more sophisticated algorithms and indicators. In this case study the efficiency of algorithm has been studied in face of this challenge. Two different scenarios have been considered for identification of this case study. In first scenario, the effect of statistical resampling (bootstrapping) number is investigated and appropriate number of statistical resample is examined on the case study identification process. Statistical resampling is a kind of bagging method of ensemble learning. In second scenario, using a stacking method of ensemble learning, several different evidences have been used in the modal identification algorithm instead of a single evidence and the results of the identification algorithm have been compared with the first scenario. Identification of close-modes systems is of great importance in modal identification viewpoint. Moreover, there is evidence that numbers of samples greater than 100 lead to negligible improvements in the estimation of standard errors [69]. In fact, according to the original developer of the bootstrapping method, even setting the number of samples at 50 is likely to lead to fairly good standard error estimates [70]. Case study is a cooling fan that has several blades. Figure 11-a and 11-b, shows the upper view and the specimen location on the diesel engine.

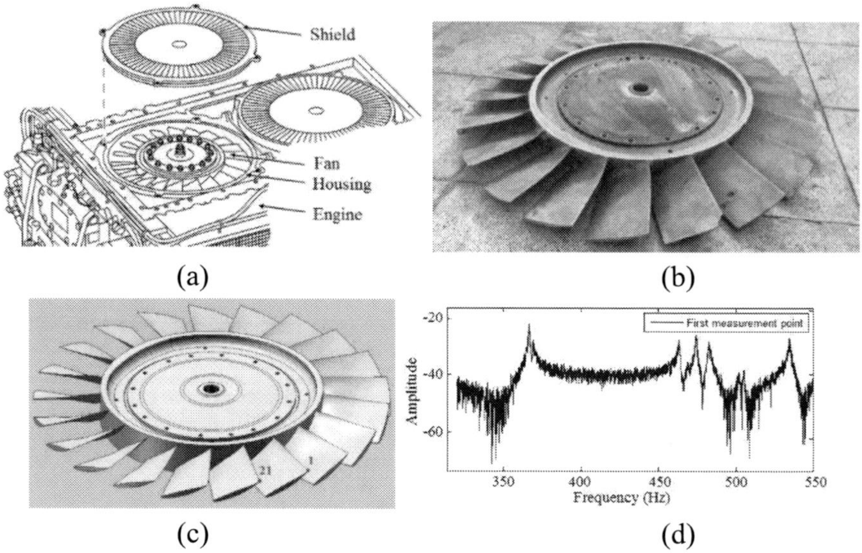

Figure 11. (a) The fan position on the engine, (b) the fan be used for operational modal experiment, (c) the upper view and measurement points, (d) the measured vibrational signal in frequency domain.

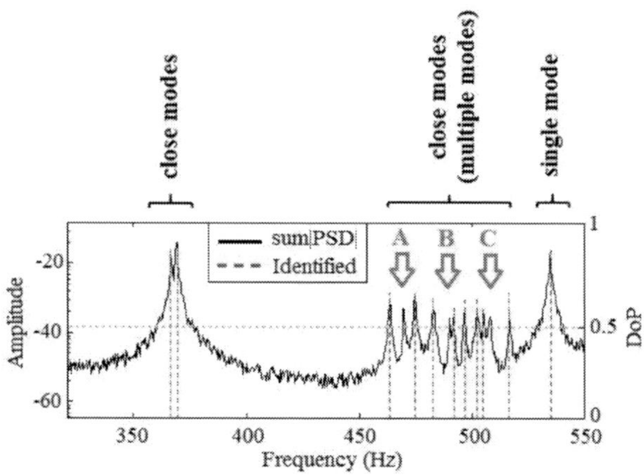

Figure 12. Identified physical modes for 2100 statistical signal.

Figure 11-c shows the measurement points. Figure 11-d, shows a measured vibrational signal of initial dataset which is measured of the first measurement point. The range of 330 to 550 Hz has been investigated and data measurement has been done at tip of each blade for 2 second by sampling frequency 10 kHz, these vibrational signals make initial dataset. 100 statistical signals are generated for each signal in initial dataset based on suggested resampling number in [69], and only one evidence (FCM) has been used to separate physical modes from spurious modes. With this number of resampling, the number of signals in the initial dataset increases from 21 signals to 2100 signals, then ensemble modal identification process is performed and the identified modes are presented in Figure 12.

The mean (average) of the PSDs is represented by a black line and the identified modes are represented by the vertical dashed red lines. But the identification algorithm only succeeded in identifying 11 vibrational modes and three modes have not been identified, as you can see in the Figure 11, the three modes (A, B, C) are not identified. It can be said that the modes which had been identified were almost identical in all identification process. Modal identification using other proposed number for statistical resampling such as reference [70], has also not been successful and it is observed that by reducing the number of resamples, fewer modes are successfully identified. The appropriate number of resamples will be obtained by an evolutionary optimization algorithm that called Imperialist Competitive Algorithm (ICA) [71]. Once the optimization procedure of resampling number is complete, the minimum resampling number is presented. This is equal to 137 resamples per dataset member. Using this sampling number, the signals of the initial dataset increase from 21 signals to 2877. This resample number is the appropriate resample number which is used in first scenario. The results of the identification algorithm by the new statistical dataset is shown in Figure 13-a. The horizontal axis is the frequency range and the vertical axis is vibrational amplitude in dB scale. Degree of Physicalness (DoP) is also shown in unit scale. In second scenario, statistical resampling was not performed based on the appropriate resampling number, the number of resampling was based on the reference [69], but several evidences were used. This means that based on the stacking method in the ensemble learning approach, several evidences, including FCM, DBSCAN, AGNES, LIDA, KNN simultaneously examine the features space and the decision of the evidences are fusioned by the DST. The modal identification result of the second scenario is also presented in the Figure 13-b.

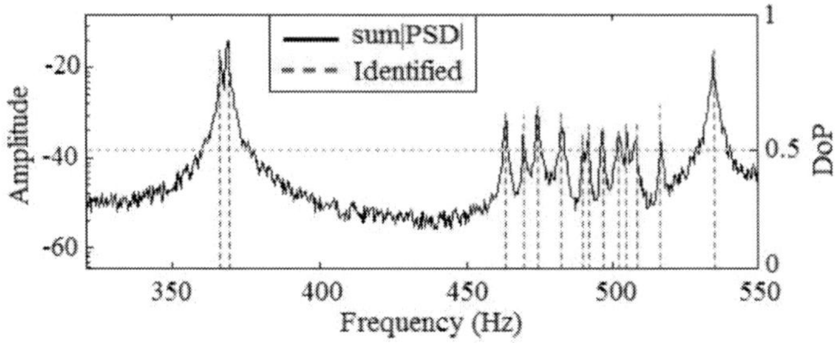

(a)

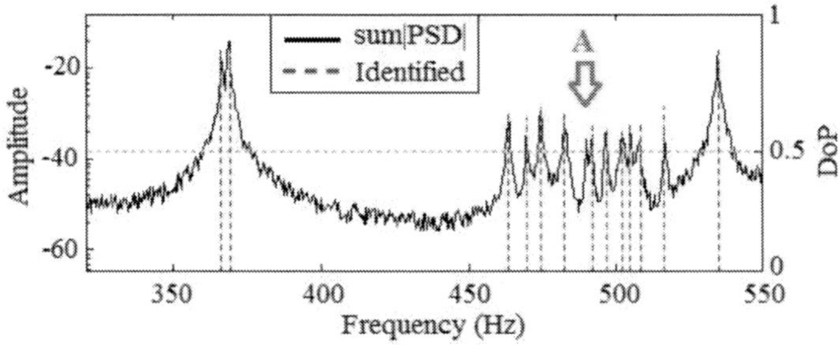

(b)

Figure 13. Identified physical modes, (a) first scenario (2877 statistical signal), (b) second scenario.

As the results show, both scenarios have increased the accuracy and efficiency of the modal identification algorithm and more modes have been identified. Using the bagging in the first scenario, all the desired modes have been identified, and using the stacking in the second scenario, only one mode has not been identified successfully. The value of identified physical modes has been listed in Table 3. The values shown in red color have been identified after changing the number of statistical resamples to the optimum value.

By increasing the number of statistical signals to appropriate value, the identification algorithm has more ability to identify less excited modes which are present in some member of dataset or modes that have less observability. This means that in the field of system identification and in the face of close modes, the ensemble learning approach is very useful and increases efficiency of identification algorithm.

Table 3. Identified mode by ensemble learning approach

Mode no.	**1**	**2**	**3**	**4**	**5**	**6**	**7**	**8**	**9**	**10**	**11**	**12**	**13**	**14**
Exp (Hz)	365.4	370.1	461.9	470.8	475	484.7	490.4	492.1	495.9	503.2	405.8	509.2	518.3	532.1
IA (Hz)	366.5	369.3	463.3	469.3	474.5	482.5	489.7	491.7	496.5	502.1	504.9	507.7	516.5	534.5

CONCLUSION

In this chapter, an ensemble learning approach in modal identification was studied. An extensive validation was carried out based on experimental results those independently obtained by using well-established experimental techniques. As an estimate we can say: “Ensemble learning is useful in system identification and modal identification is also fully performed in presence of close modes if the resampling number is assumed 40% higher than the suggested resampling number such as Efron et al. [69] and Goodhue et al. [70]”. This may potentially lead to a standardized, extensive characterization of resampling number in statistics, which is useful to gain knowledge to develop predictive identification methods.

ACKNOWLEDGMENTS

The authors would like to thank the Modal Analysis Lab of Isfahan University of Technology (IUT) for their technical assistance in this research.

REFERENCES

[1] Kocan, C. (2020). A comparative study on in-flight modal identification of an aircraft using time and frequency-domain techniques. *Journal of Vibration and Control*.

[2] Ni, Y. C. & Zhang, F. L. (2019). Fast Bayesian frequency domain modal identification from seismic response data. *Computer & Structer*, *212*, 225-235.

[3] Li, X., Antoni, J., Brennan, M. J., Yang, T. & Liu, Z. (2020). A frequency domain blind identification method for operational modal analysis using a limited number of sensors. *Journal of Vibration and Control*.

[4] McNeill, S. I. (2011). An analytic formulation for blind modal identification," *Journal of Vibration and Control*, *18* (14), 2111-2121.

[5] Peeters, B. (2000). System Identification and Damage Detection in Civil Engineering," PhD thesis, kuLeuven.

[6] Verboven, P., Cauberghe, B. & Guillaume, P. (2005). Improved total least squares estimators for modal analysis. *Computer & Structure*, *83*, 2077-2085.

[7] Li, W., Vu, V. H., Liu, Z., Thomas, M. & Hazel, B. (2017). Extraction of modal parameters for identification of time-varying systems using data-driven stochastic subspace identification. *Journal of Vibration and Control*, *24* (20), 4781-4796.

[8] Lin, R. M. (2010). Development of a new and effective modal identification method-mathematical formulations and numerical simulations. *Journal of Vibration and Control*, *17* (5), 741-758.

[9] McKelvey, T., Akcay, H. & Ljung, L. (1996). Subspace-Based Multivariable System Identification from Frequency Response Data. *IEEE Transactions on Automatic Control*, *41* (7), 960-979.

[10] Neu, E., Janser, F., Khatibi, A. A. & Orifici, A. C. (2017). Fully automated operational modal analysis using multi-stage clustering. *Mechanical Systems and Signal Processing*, *84*, 308-323.

[11] Zonno, G., Aguilar, Rafael., Castañeda B., Boroschek, R. & Lourenço, P. B. (2017). Laboratory evaluation of a fully automatic modal identification algorithm using automatic hierarchical clustering approach. *Procedia Engineering*, *199*, 882-887.

[12] Yaghoubi, V., Vakilzadeh, M. K. & Abrahamsson, T. J. S. (2017). Automated Modal Parameter Estimation Using Correlation Analysis and Bootstrap Sampling. *Mechanical Systems and Signal Processing*, *100*, 289-310.

[13] Faure, C., Ablitzer, F., Antoni, J. & Pézerat, C. (2017). Empirical and fully Bayesian approaches for the identification of vibration sources from transverse displacement measurements. *Mechanical Systems and Signal Processing*, *94*, 180-201.

[14] Nasser, F., Li, Z. y., Martin, N. & Gueguen, P. (2016). An automatic approach towards modal parameter estimation for high-rise buildings

of multicomponent signals under ambient excitations via filter-free random decrement technique. *Mechanical Systems and Signal Processing*, *70*, 821-831.

[15] Shah, M. A. S. A., Yunus, M. A., Rani, M. N. A., Tormodi, N. A., Isa, A. A. M. & Mirza, W. I. I. W. I. (2018). Closed Mode Separation of the Symmetrical Structure. *Journal of Mechanical Engineering*, *5* (3), 88-97.

[16] Walther, H., Kmetyk, L., Holzmann, W. & Segalman, D. (2004). Model Correlation with Closely Spaced Modes. *22 nd IMAC Conference & Exposition*, January 26-29, Dearborn, Michigan, USA.

[17] Green, P. L. & Maskell, S. (2017) Estimating the parameters of dynamical systems from Big Data using Sequential Monte Carlo samplers. *Mechanical Systems and Signal Processing*, *93*, 379-396.

[18] Tondreau, G. & Deraemaeker, A. (2013) Local modal filters for automated data-based damage localization using ambient vibrations. *Mechanical Systems and Signal Processing*, *39*, 162-180.

[19] Basseville, M., Abdelghani, M. & Benveniste, A. (2000). Subspace-based fault detection algorithms for vibration monitoring. *Automatica*, *36*, 101-109.

[20] Tarinejad, R. & Pourgholi, M. (2018). Modal identification of arch dams using balanced stochastic subspace identification. *Journal of Vibration and Control*, *24* (10), 2030-2044.

[21] Andersen, P., Brincker, R., Goursat, M. & Mevel, L. Automated Modal Parameter Estimation For Operational Modal Analysis of Large Systems. *Proceedings of the 2nd International Operational Modal Analysis Conference, IOMAC*, Vol. *1*, pp. (299-308).

[22] Rainieri, C. & Fabbrocino, G. (2015). Development and validation of an automated operational modal analysis algorithm for vibration-based monitoring and tensile load estimation. *Mechanical Systems and Signal Processing*, *60*, 512-534.

[23] Rainieri, C. & Fabbrocino, G. (2013). Accurate Damping Estimation by Automated OMA Procedures. *Topics in Dynamics of Civil Structures*, *4*, 1-9.

[24] Martarelli, M., Revel, G. M. & Santolini, C. (2001). Automated Modal Analysis by Scanning Laser Vibrometry: Problems and Uncestainties Associated With the Scanning System Calibration. *Mechanical Systems and Signal Processing*, *15* (3), 581-601.

[25] Bittanti, S. & Lovera, M. (2000). Bootstrap-based estimates of uncertainty in subspace identification methods. *Automatica.*, *36*, 1605-1615.

[26] Bayraktar, A., Altunişik, A. C., Sevim, B. & Türker, T. (2010). Seismic Response of a Historical Masonry Minaret using a Finite Element Model Updated with Operational Modal Testing. *Journal of Vibration and Control.*, *17* (1), 129-149.

[27] Torres, W., Almazán, J. L., Sandoval, C. & Boroschek, R. (2017). Operational modal analysis and FE model updating of the Metropolitan Cathedral of Santiago, Chile. *Engineering Structures.*, *143*, 169-188.

[28] Rainieri, C., Fabbrocino, G. & Cosenza, E. (2010). Structural Health Monitoring through automated OMA techniques in operation and during seismic events. *Conference: Proceedings of the 14th European Conference on Earthquake Engineering*.

[29] Guo, H. Y. & Li, Z. L. (2013). Structural damage identification based on evidence fusion and improved particle swarm optimization. *Journal of Vibration and Control.*, *20* (9), 1279-1292.

[30] Rainieri, C., Fabbrocino, G. & Cosenza, E. An Approach to Automated Modal Parameter Identification for Structural Health Monitoring Applications. *The Ninth International Conference on Computational Structures Technology*.

[31] Ubertini, F., Gentile, C. & Materazzi, A. L. (2013). Automated modal identification in operational conditions and its application to bridges. *Engineering Structures.*, *46*, 264-278.

[32] Vu, V. H., Thomas, M., Lafleur, F. & Marcouiller, L. (2013). Towards an automatic spectral and modal identification from operational modal analysis. *Journal of Sound and Vibration.*, *332*, 213-227.

[33] Woźniak, M., Graña, M. & Corchado, E. (2014). A survey of multiple classifier systems as hybrid systems. *Information Fusion.*, *16*, 3-17.

[34] Breiman, L. (1996). Bagging predictors. *Machine learning.*, *24* (2), 123-140.

[35] Freund, Y. & Eschapire, R. (1995). A desicion-theoretic generalization of on-line learning and an application to boosting," *Computational learning theory.*, *55* (1), 119-139.

[36] Duin, R. P. W. (2002). The combining classifier: to train or not to train?. *Pattern Recognition, Proceedings. 16th International Conference.*

[37] Britto, A. S. J., Sabourin, R. & Oliveira, L. E. S. (2014). Dynamic selection of classifiers-a comprehensive review. *Pattern Recognition.*, *47* (11), 3665-3680.

[38] Cruz, R. M. O., Sabourin, R., Cavalcanti, G. D. C. & Ren, T. (2015). META-DES: A dynamic ensemble selection framework using meta-learning. *Pattern Recognition.*, *48* (5), 1925-1935.

[39] Mu, X., Lu, J., Watta, P. & Hassoun, M. H. (2009). Weighted voting-based ensemble classifiers with application to human face recognition and voice recognition. International Joint Conference on Neural Networks.

[40] Boyun, Z., Jianping, Y., Jingbo, H., Dingxing, Z. & Shulin, W. (2007). Malicious Codes Detection Based on Ensemble Learning. Autonomic and Trusted Computing. *Lecture Notes in Computer Science.*, *4610*, 468-477.

[41] Menahem, E., Shabtai, A., Rokach, L. & Elovici, Y. (2009). Improving malware detection by applying multi-inducer ensemble". *Computational Statistics & Data Analysis.*, *53*, (4), 1483–1494.

[42] Yu, S., Shan, S., Chen, X. & Gao, W. (2006). Hierarchical ensemble of Gabor Fisher classifier for face recognition. Automatic Face and Gesture Recognition. *7th International Conference on Automatic Face and Gesture Recognition*. 91–96.

[43] Su. Y., Shan, S., Chen, X. & Gao, W. (2006). Patch-based gabor fisher classifier for face recognition," *Proceedings - International Conference on Pattern Recognition.*, *2*, 528-531.

[44] Liu, Y., Lin, Y. & Chen, Y. (2008). Ensemble Classification Based on ICA for Face Recognition. *Proceedings - 1st International Congress on Image and Signal Processing, IEEE Conference.*, 144–148.

[45] Rieger, S. A., Muraleedharan, R. & Ramachandran, R. P. (2014). Speech based emotion recognition using spectral feature extraction and an ensemble of kNN classifiers. *Proceedings of the 9th International Symposium on Chinese Spoken Language Processing.*, 589–593.

[46] Krajewski, J., Batliner, A. & Kessel, S. (2010). Comparing Multiple Classifiers for Speech-Based Detection of Self-Confidence -A Pilot Study. *20th International Conference on Pattern Recognition.*, 3716–3719.

[47] Rani P. I. & Muneeswaran, K. (2016). Recognize the facial emotion in video sequences using eye and mouth temporal Gabor features. *Multimedia Tools and Applications.*, *76* (7), 10017–10040.

[48] Rani P. I. & Muneeswaran, K. (2016). Facial Emotion Recognition Based on Eye and Mouth Regions". *International Journal of Pattern Recognition and Artificial Intelligence.*, *30* (7).

[49] Rani, P. I. & Muneeswaran, K. (2018). Emotion recognition based on facial components. *Sādhanā*. 43 (3).

[50] Louzada, F. & Ara, A. (2012). Bagging k-dependence probabilistic networks: An alternative powerful fraud detection tool" *Expert Systems with Applications.*, *39* (14), 11583–11592.

[51] Sundarkumar, G. G. & Ravi, V. (2015). A novel hybrid undersampling method for mining unbalanced datasets in banking and insurance" *Engineering Applications of Artificial Intelligence.*, *37*, 368–377.

[52] Kim, Y. & Sohn, S. Y. (2012). Stock fraud detection using peer group analysis. *Expert Systems with Applications.*, *39* (10), 8986–8992.

[53] Kumar, A. P. R. & Selvakumar, S. (2011). Distributed denial of service attack detection using an ensemble of neural classifier. *Computer Communications.*, *34* (11), 1328–1341.

[54] Savio, A., García-Sebastián, M. T., Chyzyk, D., Hernandez, C., Graña, M., Sistiaga, A., Munain, A. L. & Villanúa, J. (2011). Neurocognitive disorder detection based on feature vectors extracted from VBM

analysis of structural MRI. *Computers in Biology and Medicine.*, *41* (8), 600–610.

[55] Ayerdi, B., Savio, A. & Graña, M. (2013). Meta-ensembles of classifiers for Alzheimer's disease detection using independent ROI features. *Lecture Notes in Computer Science.*, *7931*, 122–130.

[56] Gu, Q., Ding, Y. S. & Zhang, T. L. (2015). An ensemble classifier based prediction of G-protein-coupled receptor classes in low homology. *Neurocomputing*, *154*, 110–118.

[57] Mirsadeghi, L., Kavousi, K., Hajihosseini, R. & Moghaddam, A. M. B. Evaluation of Ensemble Classifier (EC) Machine Learning Methods for Introduction of Breast Cancer Genomic Biomarkers. *Multidisciplinary Cancer Investigation*.

[58] Nath, A. & Sahu, G. K. (2019). Exploiting ensemble learning to improve prediction of phospholipidosis inducing potential. *J Theor Biol.*, *479*, 37-47.

[59] Rokach, L. (2010). Ensemble-based classifiers. *Artificial Intelligence Review.*, *33*, (1–2), 1–39.

[60] Wolpert, D. H. & Macready, W. G. (1999). An Efficient Method to Estimate Bagging's Generalization Error. *Machine Learning.*, *35*, 41-55.

[61] Clarke, B. (2003). Comparing Bayes model averaging and stacking when model approximation error cannot be ignored. *Journal of Machine Learning Research.*, 683-712.

[62] Shafer, G. (1976). A Mathematical Theory of Evidence. Princeton University Press.

[63] George, T. & Pal, N. R. (1996). Quantification of conflict in Dempster–Shafer framework: a new approach. *Int J Gen Syst*, *24* (4), 407–423.

[64] Brincker, R. & Andersen, P. (2006). Understanding Stochastic Subspace Identification. IMAC-XXIV: A Conference & Exposition on Structural Dynamics Society for Experimental Mechanics.

[65] Zhang, G., Tang, B. & Tang, G. (2012). An improved stochastic subspace identification for operational modal analysis. *Measurement*, *45* (5), 1246-1256.

[66] Overschee, P. V. & Moor, B. L. D. (1993). Subspace algorithms for the stochastic identification problem. *Automatica*, *29* (3), 649-660.

[67] Yaghoubi, V. & Abrahamsson, T. (2014). The modal observability correlation as a modal correlation metric. *Topics in Modal Analysis*, *7*, 487-494.

[68] Hall, J. F. (1988). The dynamic and earthquake behaviour of concrete dams: review of experimental behaviour and observational evidence. Elsevier.

[69] Goodhue, D. L., Lewis, W. & Thompson, R. (2012). Does PLS have advantages for small sample size or non-normal data?. *MIS Quarterly*, *36* (3), 981-1000.

[70] Efron, B., Rogosa, D. & Tibshirani, R. (2004). Resampling methods of estimation. In N.J. Smelser, & P.B. Baltes (Eds.). In book: *International Encyclopedia of the Social & Behavioral Sciences*, 13216-13220.

[71] Atashpaz, E. & Lucas, C. (2007). Imperialist competitive algorithm: an algorithm for optimization inspired by imperialistic competition. *IEEE congress on evolutionary computation*, 4661-4667.

Biographical Sketch

Dr. Mahdi Ghamami received his bachelor degree in mechanical engineering in 2009 from the Zanjan University (Zanjan, Iran), an M.S. degree in mechanical engineering in 2012 from Tehran University (Tehran, Iran), and a PhD degree in mechanical engineering in 2020 from Isfahan University of Technology (Isfahan, Iran). Dr. Ghamami has more than 8 years' academic and industrial experience in machine learning, artificial intelligence and data mining. His current research focuses on ensemble learning, system identification and fault detection in mechanical systems.

About the Editor

Yi-Tung Chan, PhD
Department of Electrical Engineering,
R. O. C. Naval Academy, Kaohsiung, Taiwan
Email: eadown92@gmail.com

Yi-Tung Chan received an Associate's degree in electronic communications engineering in 2003 from the R.O.C. Naval Academy, (Kaohsiung, Taiwan), an MS degree in computer science in 2010 from Chung Cheng Institute of Technology at the National Defense University (Taiwan), and a PhD degree from the School of Defense Science at the Chung Cheng Institute of Technology of the National Defense University (Taiwan).

Dr. Chan graduated summa cum laude from the R.O.C. Naval Academy, Chung Cheng Institute of Technology, and School of Defense Science in of 2003, 2010, and 2017, respectively.

Dr. Chan has more than 12 years' experience in research and work with ensemble learning. He was a recipient of the Academic Research Award for Excellence from the R.O.C. Naval Academy in 2018 and 2020, respectively. He also obtained the R.O.C. Naval Outstanding Teacher Award in 2020. His current research focuses on ensemble learning, maritime foreground detection, abnormal behavior detection, intelligent maritime surveillance, deep learning, machine learning, and artificial intelligence, edge computing.

Index

E

F

G

H

K

L